Carbon Nanomaterials for Therapy, Diagnosis, and Biosensing

Carbon Nanomaterials for Therapy, Diagnosis, and Biosensing

Editors

Antonino Mazzaglia
Anna Piperno

MDPI • Basel • Beijing • Wuhan • Barcelona • Belgrade • Manchester • Tokyo • Cluj • Tianjin

Editors
Antonino Mazzaglia
University of Messina
Italy

Anna Piperno
University of Messina
Italy

Editorial Office
MDPI
St. Alban-Anlage 66
4052 Basel, Switzerland

This is a reprint of articles from the Special Issue published online in the open access journal *Nanomaterials* (ISSN 2079-4991) (available at: https://www.mdpi.com/journal/nanomaterials/special_issues/carbon_nano_biosen).

For citation purposes, cite each article independently as indicated on the article page online and as indicated below:

LastName, A.A.; LastName, B.B.; LastName, C.C. Article Title. *Journal Name* **Year**, *Volume Number*, Page Range.

ISBN 978-3-0365-4511-0 (Hbk)
ISBN 978-3-0365-4512-7 (PDF)

© 2022 by the authors. Articles in this book are Open Access and distributed under the Creative Commons Attribution (CC BY) license, which allows users to download, copy and build upon published articles, as long as the author and publisher are properly credited, which ensures maximum dissemination and a wider impact of our publications.

The book as a whole is distributed by MDPI under the terms and conditions of the Creative Commons license CC BY-NC-ND.

Contents

About the Editors vii

Antonino Mazzaglia and Anna Piperno
Carbon Nanomaterials for Therapy, Diagnosis and Biosensing
Reprinted from: *Nanomaterials* **2022**, *12*, 1597, doi:10.3390/nano12091597 1

Mariachiara Trapani, Antonino Mazzaglia, Anna Piperno, Annalaura Cordaro,
Roberto Zagami, Maria Angela Castriciano, Andrea Romeo and Luigi Monsù Scolaro
Novel Nanohybrids Based on Supramolecular Assemblies of Meso-tetrakis-(4-sulfonatophenyl) Porphyrin J-aggregates and Amine-Functionalized Carbon Nanotubes
Reprinted from: *Nanomaterials* **2020**, *10*, 669, doi:10.3390/nano10040669 5

Giulia Neri, Enza Fazio, Placido Giuseppe Mineo, Angela Scala and Anna Piperno
SERS Sensing Properties of New Graphene/Gold Nanocomposite
Reprinted from: *Nanomaterials* **2019**, *9*, 1236, doi:10.3390/nano9091236 19

Chiara Pennetta, Giuseppe Floresta, Adriana Carol Eleonora Graziano, Venera Cardile,
Lucia Rubino, Maurizio Galimberti, Antonio Rescifina and Vincenzina Barbera
Functionalization of Single and Multi-Walled Carbon Nanotubes with Polypropylene Glycol Decorated Pyrrole for the Development of Doxorubicin Nano-Conveyors for Cancer Drug Delivery
Reprinted from: *Nanomaterials* **2020**, *10*, 1073, doi:10.3390/nano10061073 33

Rosa Garriga, Tania Herrero-Continente, Miguel Palos, Vicente L. Cebolla, Jesús Osada,
Edgar Muñoz and María Jesús Rodríguez-Yoldi
Toxicity of Carbon Nanomaterials and Their Potential Application as Drug Delivery Systems: In Vitro Studies in Caco-2 and MCF-7 Cell Lines
Reprinted from: *Nanomaterials* **2020**, *10*, 1617, doi:10.3390/nano10081617 55

Sandra Claveau, Émilie Nehlig, Sébastien Garcia-Argote, Sophie Feuillastre,
Grégory Pieters, Hugues A. Girard, Jean-Charles Arnault, François Treussart
and Jean-Rémi Bertrand
Delivery of siRNA to Ewing Sarcoma Tumor Xenografted on Mice, Using Hydrogenated Detonation Nanodiamonds: Treatment Efficacy and Tissue Distribution
Reprinted from: *Nanomaterials* **2020**, *10*, 553, doi:10.3390/nano10030553 77

Tzu-Yin Lee, Thanasekaran Jayakumar, Pounraj Thanasekaran, King-Chuen Lin,
Hui-Min Chen, Pitchaimani Veerakumar and Joen-Rong Sheu
Carbon Dot Nanoparticles Exert Inhibitory Effects on Human Platelets and Reduce Mortality in Mice with Acute Pulmonary Thromboembolism
Reprinted from: *Nanomaterials* **2020**, *10*, 1254, doi:10.3390/nano10071254 91

Roberta Di Carlo, Susi Zara, Alessia Ventrella, Gabriella Siani, Tatiana Da Ros,
Giovanna Iezzi, Amelia Cataldi and Antonella Fontana
Covalent Decoration of Cortical Membranes with Graphene Oxide as a Substrate for Dental Pulp Stem Cells
Reprinted from: *Nanomaterials* **2019**, *9*, 604, doi:10.3390/nano9040604 107

Angelo Nicosia, Fabiana Vento, Anna Lucia Pellegrino, Vaclav Ranc, Anna Piperno,
Antonino Mazzaglia and Placido Mineo
Polymer-Based Graphene Derivatives and Microwave-Assisted Silver Nanoparticles Decoration as a Potential Antibacterial Agent
Reprinted from: *Nanomaterials* **2020**, *10*, 2269, doi:10.3390/nano10112269 121

Annalaura Cordaro, Giulia Neri, Maria Teresa Sciortino, Angela Scala and Anna Piperno
Graphene-Based Strategies in Liquid Biopsy and in Viral Diseases Diagnosis
Reprinted from: *Nanomaterials* **2020**, *10*, 1014, doi:10.3390/nano10061014 . . . . **139**

About the Editors

Antonino Mazzaglia

Antonino Mazzaglia is the Research Director at the National Council of Research-Institute for the Nanostructured Materials (CNR-ISMN) at the University of Messina (Italy). He is a member of the Chemical Science PhD Board at the University of Messina. He joined the International, European, and National Advisory Boards on Cyclodextrins and the Italian Society of Photobiology. He was the Chair of the International Summer School on Cyclodextrins (2013) and the Bridge Virtual Meeting (2021) and was the Chair of the XX International Cyclodextrin Symposium (20th ICS). He has been a visiting scientist and professor in Europe (University of Exeter, Kings College London, University of Normandie, etc.) and in various Italian Universities. He has been an invited plenary and key-note speaker at several conferences (ICS, ESP; ICPP, etc.), a Member of the Italian Research Evaluation Board (VQR) for chemistry and an Editorial Board Member of *Nanomaterials* and *IJMS*. He is the author 103 peer-reviewed papers, 7 book chapters, and 1 monographic review, accomplishing an H-index of 30 and 2336 citations (SCOPUS source). His research expertise is focused on bio-soft hybrid materials based on cyclodextrins, porphyrinoids and photosensitisers, polysaccharides, polymers, and carbon nanoplatforms that have stimuli-responsiveness properties for therapy, diagnosis, and biosensing.

Anna Piperno

Anna Piperno earned a Master's degree in Pharmaceutical Chemistry and Technology in 1994 and a PhD in Pharmaceutical Science in 1999 at the University of Messina (Italy). She obtained a permanent position at Messina University as a researcher in 2001 and is currently a full professor in organic chemistry at the University of Messina. Since 2001, she has carried out teaching activities in organic chemistry and has supervised several students and doctoral fellows. She is a member of the Chemical Science PhD Board at the University of Messina. She began her research activities by studying organic synthesis methodologies and the chemistry of heterocyclic compounds. Specifically, her work was focused on the design and synthesis of new compounds that interfere with viral replication or cell death/proliferation. Successively, her research interests have been extended to the functionalization of carbon nanomaterials and biopolymers for applications in drug delivery, regenerative medicine, and biosensing. She actively collaborates with pharmaceutical industries on projects in the field of drug delivery and liquid biopsy. Currently, she is the PI of the MSCA Doctoral Networks "STRIKE" project and of NATO's "VIPER" project. She is the author of 123 papers that have been published in peer-reviewed journals and book chapters (H index = 36, citations = 3143, Scopus).

 *nanomaterials*

Editorial

Carbon Nanomaterials for Therapy, Diagnosis and Biosensing

Antonino Mazzaglia [1,*] and Anna Piperno [2,*]

1. CNR-ISMN, Istituto per lo Studio dei Materiali Nanostrutturati, URT of Messina, Department of Chemical, Biological, Pharmaceutical and Environmental Sciences, University of Messina, V. le F. Stagno d'Alcontres 31, 98166 Messina, Italy
2. Department of Chemical, Biological, Pharmaceutical and Environmental Sciences, University of Messina, V. le F. Stagno d'Alcontres 31, 98166 Messina, Italy
* Correspondence: antonino.mazzaglia@cnr.it (A.M.); anna.piperno@unime.it or apiperno@unime.it (A.P.)

Citation: Mazzaglia, A.; Piperno, A. Carbon Nanomaterials for Therapy, Diagnosis and Biosensing. *Nanomaterials* 2022, 12, 1597. https://doi.org/10.3390/nano12091597

Received: 23 April 2022
Accepted: 27 April 2022
Published: 9 May 2022

Publisher's Note: MDPI stays neutral with regard to jurisdictional claims in published maps and institutional affiliations.

Copyright: © 2022 by the authors. Licensee MDPI, Basel, Switzerland. This article is an open access article distributed under the terms and conditions of the Creative Commons Attribution (CC BY) license (https://creativecommons.org/licenses/by/4.0/).

In carbon nanomaterial design, the fine-tuning of their functionalities and physicochemical properties has increased their potential for therapeutic, diagnostic and biosensing applications [1–3]. In this Special Issue, articles or mini reviews on nanoplatforms originating from the synergistic combination of carbon-based nanomaterials (i.e., nanotubes, graphene, graphene oxide, carbon quantum dots, nanodiamond, etc.) and various functional molecules such as drugs, natural compounds, biomolecules, polymers, metal nanoparticles and macrocycles relevant in drug delivery, in multi-targeted therapy, in theragnostics, as scaffolds in tissue engineering and as a sensing material, have been selected for publication.

Trapani et al. investigated the ability of multiwalled carbon nanotubes (MWCNTs) covalently modified with polyamine chains of various length (ethylenediamine (EDA) and tetraethylenepentamine (EPA)) to induce the J-aggregation of meso-tetrakis (4-sulfonatophenyl)porphyrin (TPPS) in different experimental conditions. The authors reported that, in mild acidic conditions, TPPS porphyrin easily self-assembles into J-aggregates, showing peculiar extinction bands in the visible region ($\lambda \cong 493$ nm) and in the therapeutic window ($\lambda \cong 710$ nm), together with an emission band in the red spectral region. The results of this study describe the experimental conditions in which to obtain stable TPPS J-aggregates in medium mimicking physiological conditions for a stimuli-responsive therapeutic action upon irradiation on their extinction bands and fluorescence probes in cellular environments [4].

In the design of diagnostic nanoplatforms, aiming to improve the surface-enhanced Raman spectroscopy (SERS) effect, Neri et al. proposed a new graphene/gold nanocomposite composed of gold nanoparticles (AuNPs), produced by pulsed laser ablation in liquids (PLAL), and a nitrogen-doped graphene platform (G-NH2) obtained by direct delamination and chemical functionalization of graphite flakes with 4-methyl-2-p-nitrophenyl oxazolone, followed by the reduction of p-nitrophenyl groups. This approach allowed the authors to study SERS properties of graphene loaded with pure AuNPs without the influence of capping agents, surfactants, or salt produced in the chemical reduction of gold ions. The SERS platform was tested for its ability to detect Rhodamine 6G and Dopamine as molecular probes at a concentration around 1 µM. The platform showed good stability and the ability to reproduce Raman signals without degradation although its sensitivity was low. Overall, the feasibility of the proposed method opens up the field to further research on improving the detection limits of molecular probes interacting with loaded AuNPs [5].

Nowadays, new therapeutic approaches using carbon nanomaterials have become very attractive. In this scenario, Pennetta et al. investigated the formation of Doxorubicin (DOX) nano-conveyors as a stacked drug-delivery system for application in cancer treatment. The innovative nanoplatform was obtained by functionalizing single- and multi-walled carbon nanotubes (CNT and MWCNT, respectively) by cycloaddition reaction between carbon nanotubes and a pyrrole-derived compound. Two different adducts between CNT and pyrrole polypropylene glycol (PPGP) were prepared: the supramolecular adduct (CNT/PPGP$_s$) and the covalent one (CNT/PPGP$_c$). The supramolecular interactions were studied on the basis of molecular dynamics simulations, and by monitoring the emission and the absorption spectra of DOX. The biological studies revealed that two of the synthesized nanoplatforms are effectively able to obtain DOX within A549 and M14 cell lines and to enhance the cell mortality at a much lower effective dose of DOX. This work paves the way for the facile functionalization of carbon nanotubes by exploiting the "pyrrole methodology" for the development of novel technological carbon-based drug-delivery systems [6].

One of the main widespread concerns regarding using carbon-based materials as alternative nanobiomaterials for cancer therapy is their inherent cytotoxicity, which remains debated, with studies demonstrating contradictory results. Garriga et al. investigated the in vitro toxicity of various carbon nanomaterials in human epithelial colorectal adenocarcinoma (Caco-2) cells and human breast adenocarcinoma (MCF-7) cells. Carbon nanohorns (CNH), carbon nanotubes (CNT), carbon nanoplatelets (CNP), graphene oxide (GO), reduced graphene oxide (GO) and nanodiamonds (ND) were systematically evaluated and compared using Pluronic F-127 as a dispersant agent. Carbon nanomaterial exposure affected the cell viability in the following order: CNP < CNH < RGO < CNT < GO < ND, with a pronounced effect on the more rapidly dividing Caco-2 cells. Hydrophobicity and morphological features are the main causes of decreases in cell viability, enhanced levels of ROS (radical oxygen species) and apoptosis/necrosis. In this study, ND showed low toxicity thanks to a lack of ROS levels and was efficient in the loading of hydrophilic drugs, such as DOX, by assembling in the ND surface or within the pores. On the other hand, this study evidenced the low toxicity of CNT and RGO, and the high camptothecin (CPT) loading because of the strong π–π stacking interactions. Altogether, despite the various obstacles that still have to be overcome before considering carbon nanomaterials suitable as drug carriers (i.e., the potential long-term toxicity), this study highlighted a screening and risk-to-benefit assessment and, together with drug-loading efficiency studies, is fundamental to the development of advanced multi-functional carbon nanomaterials for cancer theragnostic applications [7].

Nanodiamonds with detonation origins were investigated by Claveau et al. as delivery systems for anti-cancer therapy in vivo models. The authors studied the ability of cationic hydrogenated detonation nanodiamonds to carry active small interfering RNA (siRNA) in a mice model of Ewing sarcoma, which is bone cancer of young adults due to the *EWS-FLI1* junction oncogene in the majority of patients. Labeled nanodiamonds obtained using radioactive tritium gas instead of hydrogen gas allowed the authors to investigate the trafficking of nanodiamonds throughout mouse organs and their excretion as urine and feces. Moreover, the ability of siRNA to inhibit the expression of the oncogene *EWS-FLI1* in tumor-xenografted mice was demonstrated. Overall, this study represents a substantial step towards the use of ultra-small solid nanoparticles for the delivery of nucleic acid in vivo [8].

In the framework of carbon nanomaterial development for therapeutic purposes, Lee et al. reported a new type of carbon dot (CDOT) nanoparticle as a new antiplatelet agent. The inhibition of platelet activation is considered a potential therapeutic approach for the treatment of arterial thrombotic diseases; therefore, maintaining platelets in their inactive state has gained much attention. CDOT could actively inhibit human platelet activation by suppressing some crucial mechanisms (e.g., PKC activation, and Akt, JNK1/2 and p38 MAPK phosphorylation) with no in vitro cytotoxicity. This in vivo study revealed that the CDOTs had an antithrombotic effect on the ADP-induced pulmonary thromboembolic mice model by reducing mortality and by preserving the normal bleeding tendency in mice. Altogether, these results suggest that a direct application of CDOTs may contribute to the development of new antiplatelet drugs for the treatment of arterial thromboembolic diseases [9].

In the application of graphene nanomaterials for dental regenerative engineering, Di Carlo et al. proposed the covalent functionalization of a graphene oxide (GO)-decorated cortical membrane (Lamina®) in the promotion of the adhesion, growth and osteogenic differentiation of DPSCs (Dental Pulp Stem Cells). The GO-decorated Laminas demonstrated an increase in the roughness of Laminas and a reduction in toxicity and did not affect the membrane integrity of DPSCs. In conclusion, this study showed that the GO covalent functionalization of Laminas was effective; was relatively easy to obtain; and favored both the proliferation rate of DPSCs, probably due to the capacity of GO to adsorb proteins present in the medium, and the deposition of calcium phosphate. Overall, this study is promising because the proposed material holds potential as a useful substrate in facilitating in vivo bone regeneration [10].

In antibacterial applications, Nicosia et al. synthesized novel NanoHybrid Systems based on graphene, polymers and AgNPs (namely, NanoHy-GPS) using an easy microwave irradiation approach free of reductants and surfactants. The fine-tuned hybrid system combines the properties of polymers, graphene and AgNPs as a potential on-demand antimicrobial coating system. Polymers play key roles in ensuring the coating compatibility of the graphene platform, making it adaptable for a specific substrate. The driving force of this strategy is the tuning of the interfacial interactions towards targeted substrates, thus optimizing the homogeneity of the dispersion of the GO derivatives within specific polymer matrices. The formulation of functionalized graphene with AgNPs entrapped in suitable polymers resulted in a doubly beneficial effect: an increase in graphene processability and achievement of graphene-enriched antimicrobial nanobiomaterials. NanoHy-GPS was proposed as a potential alternative to common antibacterial agents, which leak into the environment and/or within organisms' tissues [11].

Finally, Cordaro et al. proposed a review aimed at providing a comprehensive and exhaustive summary of the contributions of graphene-based nanomaterials to liquid biopsy. Liquid biopsy is considered an innovative method that has provided surprising perspectives in the early diagnosis of severe diseases such as cancer, metabolic syndrome, and autoimmune and neurodegenerative disorders and in monitoring their treatment. Although nanotechnology based on graphene has been poorly applied for the rapid diagnosis of viral diseases, the extraordinary features of graphene (i.e., high electronic conductivity, large specific area and surface functionalization) can also be exploited for the diagnosis of emerging viral diseases, such as coronavirus disease 2019 (COVID-19) [12].

The variety of applications covered by the nine articles published in this Special Issue of *Nanomaterials* "Carbon Nanomaterials for Therapy, Diagnosis and Biosensing" is proof of the growing attention on the use of carbon nanomaterials in the biomedical/pharmaceutical field in recent years. We hope that the readers enjoy reading these articles and find them useful for their research and for advancing carbon nanomaterials from the laboratory to clinical nanomedicine. Finally, we acknowledge all of the authors who contributed their work to this Special Issue as well as the editorial board of *Nanomaterials* for all of their support.

Funding: This research received no external funding.

Conflicts of Interest: The authors declare no conflict of interest.

References

1. Díez-Pascual, A.M. Carbon-Based Nanomaterials. *Int. J. Mol. Sci.* **2021**, *22*, 7726. [CrossRef] [PubMed]
2. Maiti, D.; Tong, X.; Mou, X.; Yang, K. Carbon-Based Nanomaterials for Biomedical Applications: A Recent Study. *Front. Pharmacol.* **2019**, *9*, 1401. [CrossRef] [PubMed]
3. Zhang, D.Y.; Zheng, Y.; Tan, C.P.; Sun, J.H.; Zhang, W.; Ji, L.N.; Mao, Z.-W. Graphene oxide decorated with Ru(II)–polyethylene glycol complex for lysosome-targeted imaging and photodynamic/photothermal therapy. *ACS Appl. Mater. Interfaces* **2017**, *9*, 6761–6771. [CrossRef] [PubMed]
4. Trapani, M.; Mazzaglia, A.; Piperno, A.; Cordaro, A.; Zagami, R.; Castriciano, M.A.; Romeo, A.; Monsù Scolaro, L. Novel Nanohybrids Based on Supramolecular Assemblies of Meso-tetrakis-(4-sulfonatophenyl) Porphyrin J-aggregates and Amine-Functionalized Carbon Nanotubes. *Nanomaterials* **2020**, *10*, 669. [CrossRef] [PubMed]

5. Neri, G.; Fazio, E.; Mineo, P.G.; Scala, A.; Piperno, A. SERS Sensing Properties of New Graphene/Gold Nanocomposite. *Nanomaterials* **2019**, *9*, 1236. [CrossRef] [PubMed]
6. Pennetta, C.; Floresta, G.; Graziano, A.C.E.; Cardile, V.; Rubino, L.; Galimberti, M.; Rescifina, A.; Barbera, V. Functionalization of Single and Multi-Walled Carbon Nanotubes with Polypropylene Glycol Decorated Pyrrole for the Development of Doxorubicin Nano-Conveyors for Cancer Drug Delivery. *Nanomaterials* **2020**, *10*, 1073. [CrossRef] [PubMed]
7. Garriga, R.; Herrero-Continente, T.; Palos, M.; Cebolla, V.L.; Osada, J.; Muñoz, E.; Rodríguez-Yoldi, M.J. Toxicity of Carbon Nanomaterials and Their Potential Application as Drug Delivery Systems: In Vitro Studies in Caco-2 and MCF-7 Cell Lines. *Nanomaterials* **2020**, *10*, 1617. [CrossRef] [PubMed]
8. Claveau, S.; Nehlig, E.; Garcia-Argote, S.; Feuillastre, S.; Pieters, G.; Girard, H.A.; Arnault, J.-C.; Treussart, F.; Bertrand, J.-R. Delivery of siRNA to Ewing Sarcoma Tumor Xenografted on Mice, Using Hydrogenated Detonation Nanodiamonds: Treatment Efficacy and Tissue Distribution. *Nanomaterials* **2020**, *10*, 553. [CrossRef] [PubMed]
9. Lee, T.-Y.; Jayakumar, T.; Thanasekaran, P.; Lin, K.-C.; Chen, H.-M.; Veerakumar, P.; Sheu, J.-R. Carbon Dot Nanoparticles Exert Inhibitory Effects on Human Platelets and Reduce Mortality in Mice with Acute Pulmonary Thromboembolism. *Nanomaterials* **2020**, *10*, 1254. [CrossRef] [PubMed]
10. Di Carlo, R.; Zara, S.; Ventrella, A.; Siani, G.; Da Ros, T.; Iezzi, G.; Cataldi, A.; Fontana, A. Covalent Decoration of Cortical Membranes with Graphene Oxide as a Substrate for Dental Pulp Stem Cells. *Nanomaterials* **2019**, *9*, 604. [CrossRef] [PubMed]
11. Nicosia, A.; Vento, F.; Lucia Pellegrino, A.L.; Ranc, V.; Piperno, A.; Mazzaglia, A.; Mineo, P. Polymer-Based Graphene Derivatives and Microwave-Assisted Silver Nanoparticles Decoration as a Potential Antibacterial Agent. *Nanomaterials* **2020**, *10*, 2269. [CrossRef] [PubMed]
12. Cordaro, A.; Neri, G.; Sciortino, M.T.; Scala, A.; Piperno, A. Graphene-Based Strategies in Liquid Biopsy and in Viral Diseases Diagnosis. *Nanomaterials* **2020**, *10*, 1014. [CrossRef] [PubMed]

Article

Novel Nanohybrids Based on Supramolecular Assemblies of Meso-tetrakis-(4-sulfonatophenyl) Porphyrin J-aggregates and Amine-Functionalized Carbon Nanotubes

Mariachiara Trapani [1], Antonino Mazzaglia [1,*], Anna Piperno [2,3], Annalaura Cordaro [1,2], Roberto Zagami [1], Maria Angela Castriciano [1,*], Andrea Romeo [1,2,4] and Luigi Monsù Scolaro [1,2,4]

[1] CNR-ISMN, Istituto per lo Studio dei Materiali Nanostrutturati c/o Dipartimento di Scienze Chimiche, Biologiche, Farmaceutiche ed Ambientali, Università di Messina, V. le F. Stagno D'Alcontres 31, 98166 Messina, Italy; mariachiara.trapani@cnr.it (M.T.); annalaura.cordaro@ismn.cnr.it (A.C.); roberto.zagami@ismn.cnr.it (R.Z.); anromeo@unime.it (A.R.); lmonsu@unime.it (L.M.S.)
[2] Dipartimento di Scienze Chimiche, Biologiche, Farmaceutiche ed Ambientali, Università di Messina, V. le F. Stagno D'Alcontres 31, 98166 Messina, Italy; apiperno@unime.it
[3] Consorzio Interuniversitario Nazionale di Ricerca in Metodologie e Processi Innovativi di Sintesi, C.I.N.M.P.I.S., Unità Operativa dell'Università di Messina, V. le F. Stagno D'Alcontres, 3198166 Messina, Italy
[4] Consorzio Interuniversitario di Ricerca in Chimica dei Metalli nei Sistemi Biologici, C.I.R.C.M.S.B, Unità Operativa dell'Università di Messina, V. le F. Stagno D'Alcontres, 31, 98166 Messina, Italy
* Correspondence: antonino.mazzaglia@cnr.it (A.M.); maria.castriciano@cnr.it (M.A.C.)

Received: 13 February 2020; Accepted: 26 March 2020; Published: 2 April 2020

Abstract: The ability of multiwalled carbon nanotubes (MWCNTs) covalently functionalized with polyamine chains of different length (ethylenediamine, EDA and tetraethylenepentamine, EPA) to induce the J-aggregation of meso-tetrakis(4-sulfonatophenyl)porphyrin (TPPS) was investigated in different experimental conditions. Under mild acidic conditions, protonated amino groups allow for the assembly by electrostatic interaction with the diacid form of TPPS, leading to hybrid nanomaterials. The presence of only one pendant amino group for a chain in EDA does not lead to any aggregation, whereas EPA (with four amine groups for chain) is effective in inducing J-aggregation using different mixing protocols. These nanohybrids have been characterized through UV/Vis extinction, fluorescence emission, resonance light scattering and circular dichroism spectroscopy. Their morphology and chemical composition have been elucidated through transmission electron microscopy (TEM) and scanning transmission electron microscopy (STEM). TEM and STEM analysis evidence single or bundles of MWCNTs in contact with TPPS J-aggregates nanotubes. The nanohybrids are quite stable for days, even in aqueous solutions mimicking physiological medium (NaCl 0.15 M). This property, together with their peculiar optical features in the therapeutic window of visible spectrum, make them potentially useful for biomedical applications.

Keywords: porphyrin; J-aggregates; carbon nanotubes; nanohybrids

---

1. Introduction

Carbon nanotubes (CNTs) are intriguing materials with applications ranging from nanotechnology-related devices (i.e., in electronics, energy storage, water treatment, as sensor/biosensor) [1] to drug/probes delivery systems for therapy and diagnosis [2,3].

Functionalized CNTs are widely used to reduce the intrinsic toxicity of "as produced" (pristine) CNTs by increasing the tolerability and the biodegradability in vivo [4,5]. Furthermore, opportunely modified and sized multiwalled CNTs (MWCNTs) are not retained in the organs and can be easily cleared by body excretion [6].

Functional nanomaterials based on CNTs were designed as therapeutic enhancers by combining CNTs with different systems, such as cyclodextrins, biomolecules or porphyrinoids [3]. Recently, some of us reported antiviral- and plasmid/delivery systems [7,8] based on properly functionalized MWCNTs, investigating also their intracellular fate. Similarly to other nanomaterials based on carbon for multi-targeted therapies and imaging [9,10], CNTs functional nanomaterials were endowed with unique properties generated by the synergic actions of components [3].

Non-covalent modification of CNTs with porphyrinoids is a well-investigated strategy to modulate the environment of chromophores, thus improving the light absorption and emission features [11], charge-transport [12] or energy transfer properties [13] in view of bio-labeling and light harvesting applications. In particular, it is well-known as π-stacking of CNTs with hydrosoluble porphyrins provides donor-acceptor complexes with efficient energy transfer [14]. Porphyrin free bases or metallo porphyrins/DNA supramolecular systems undergo strong charge transfer with semiconducting CNTs [15]. Enhanced photoconductivity has been reported for J- and H- porphyrin aggregates (head-to-tail or head-to-head molecules stacking, respectively) obtained in solution by interaction with single-walled CNTs (SWCNTs) [16] or at solid state with double-walled CNTs (DWCNTs) [17] or by decorating MWCNTs film [18]. Moreover, recently it was demonstrated that photoluminescence properties of a hybrid material assembled by formation of J-aggregates of benzo[e]indocarbocyanine (BIC) on SWCNTs can be modulated by selecting cis- or trans- isomer of the dye: the first one quenches the photoluminescence by strong interaction with CNTs, whereas the second one forms free J-aggregates characterized by photoluminescence bands of practical use in biomedical imaging [19]. Indeed, it is well known that J-aggregates feature very narrow red-shifted absorption bands, showing renewed optical, photophysical, and structural properties vs. monomer [20]. In this framework, multifunctional nanotheranostic based on J-type aggregates of cyanine [21,22], bacterio-pheophorbide [23], chlorine [24], and Bodipy [25] were proposed due to their excellent photothermal and/or NIR absorbing features for applications in imaging guided therapy (i.e., photoacoustic imaging). However, these J-type aggregates generally need to be entrapped in liposomes or dispersed in surfactants to increase their solubility and bio-availability.

Within the incoming research of composite nanomaterials, our interest has been addressed to J-aggregates of meso-tetrakis-(4-sulfonatophenyl)porphyrin (TPPS) exhibiting peculiar optical features [26–29]. Such features can be fine-tuned depending on the strategy adopted to obtain the structure, i.e., by selecting appropriate polyamines as scaffolds [30–35] or by tailoring nanomaterials (i.e., metal nanoparticles) [36–40] or triggering porphyrin J-aggregation by modulating pH and/or ionic strength [41–44]. TPPS J-aggregates, in line of principle, would not necessitate further manipulation/encapsulation to explicate their properties within cells or tissues. These aggregated species could lead to stimuli-responsive therapeutic action upon irradiation on their extinction bands, fluorescence probing in cellular environments, or refilling of the dye upon eventual J-aggregates disassembly [45,46] in biological sites [47].

Regarding the design of SWCNTs/TPPS nanohybrids, assembly between TPPS with amine-conjugated SWCNTs has been obtained in water, pointing out that the photophysical properties of porphyrin are largely influenced by length of CNTs amine chain [48] whereas TPPS J-aggregates on SWCNTs have been prepared in organic solvents [49].

Herein, we report on the formation of relatively stable TPPS J-aggregates wrapped to covalently amine-modified MWCNTs in aqueous solution. We anticipate that, in the presence of tetraethylenepentamine-functionalized MWCNTs (MWCNT-EPA) in mild acidic conditions, TPPS porphyrin easily self-assembles into J-aggregates exhibiting peculiar extinction bands in the visible region (i.e., ≅493 nm) and in the therapeutic window (i.e., ≅710 nm), together with an emission band in the red spectral region for potential phototherapeutic and/or photodiagnostic applications. Conversely, ethylenediamine-modified MWCNTs (MWCNT-EDA) do not induce J-aggregate formation due to EDA structural features vs. EPA ones. The MWCNT-EPA/TPPS J-aggregates are stable in mimicking physiological medium (NaCl = 0.15 M), thus opening the route for their potential application in dual therapeutic/diagnostic assessment.

2. Materials and Methods

The 5,10,15,20-tetrakis(4 sulfonatophenyl)porphyrin (TPPS), MWCNTs, tert-butyl 2-aminoethylcarbamate (EDABoc), tetraethylenepentamine (EPA), N-(3-dimethylaminopropyl)-N′-ethylcarbodiimide hydrochloride (EDC), 1-Hydroxybenzotriazole (HOBt), Ninhydrin test kit, other solvents and reagents were purchased from Sigma-Aldrich Chemicals (Milan, Italy). The stock porphyrin aqueous solutions were freshly prepared and their concentrations were determined using the extinction coefficient at the Soret maximum ($\varepsilon = 5.33 \times 10^5$ M^{-1}cm^{-1} at $\lambda = 414$ nm). All the reagents were used without further purification and all the solutions were prepared in dust free Milli-Q water (Merck, Darmstadt, Germany).

Carboxylated multiwalled carbon nanotubes (MWCNT-Ox) were prepared by the oxidation of MWCNTs (mean diameter 5–10 nm; average length 10–20 μm) with sulfuric acid/nitric acid (3:1 v/v, 98% and 69%) according to the protocol previously reported [50]. MWCNT-Ox (100 mg) in 25 mL of dry DMF were sonicated for 30 min, then EDC (112 mg, 0.58 mmol) and HOBt (40 mg, 0.30 mmol), were added and the black suspension was stirred at room temperature for one hour. EDABoc (55 mg, 0.24 mmol) was added and the reaction was stirred for 72 hours. Water/ethanol (1:1 mixture) was added and the crude reaction mixture was filtered under vacuum (Millipore, 0.1 μm), washed with an excess of water/ethanol and finally diethyl ether. The resulting MWCNT-EDABoc were dispersed in dioxane (10 mL) and treated with 5 mL of HCl 4 M at room temperature for 4 h. The mixture was filtered under vacuum (Millipore, 0.1 μm) and the precipitate was treated with 5 mL of triethylamine/water (1:4), thus obtaining MWCNT-EDA. This was washed several times with water/ethanol by successive bath sonication and centrifugation (8000 rpm 10 min) procedures and finally dried at 60 °C to give 60 mg of material. The amount of free amine groups on MWCNT-EDA was estimated by Kaiser test (0.22 mmol/g).

MWCNT-EPA were prepared by the coupling of MWCNT-Ox and EPA, in presence of EDC/HOBt, according to the amidation reaction procedure above described. The resulting MWCNT-EPA were washed several times with water/ethanol by successive bath sonication and centrifugation (8000 rpm 10 min) procedures and finally dried at 60 °C. Termogravimetric analysis (TGA) data indicated a weight loss of about 4.3% at 500 °C, which roughly corresponds to 0.22 mmol/g of EPA. The ninhydrin assay indicated an amount of free amino groups of 0.44 mmol/g.

Primary amine loadings were measured spectroscopically using the colorimetric Kaiser conditions [51,52]. Commercial Kaiser test kit is composed of three solutions as it follows: (a) 0.5 g/mL of phenol in absolute EtOH; (b) 2 mL of potassium cyanide 1 mM (aqueous solution) dissolved in 98 mL of pyridine; (c) 0.05 g/mL of ninhydrin in absolute EtOH. Briefly, 0.5 mg of MWCNT-EDA or MWCNT-EPA were treated in sequence with 75 μL of solution (a), 100 μL of solution (b) and 75 μL of solution (c). The dispersion was sonicated in a water bath and then was heated at 120 °C for 5 min, diluted with 4750 μL of absolute EtOH and centrifugated at 14,000 rpm. The absorbance at 570 nm of supernatant was correlated to the amount of free amine groups on MWCNTs surface (NH$_2$ loading (mmol/g)); using the following equation:

$$[free\ amines] = ([Abs] \times dilution \times 1000)/(\varepsilon \times sample\ weight \times optical\ path) \qquad (1)$$

where dilution was fixed to 5 mL, optical path was 1 cm; sample weight was 0.5 mg; extinction coefficient (ε) was 15,000 M^{-1} cm^{-1}.

Dispersions of MWCNT-EDA and MWCNT-EPA (0.43 mg/mL) were prepared in 10 mM citrate buffer by bath sonication for 20 min. For the experiments, a volume of 100 μL has been used and diluted to a final concentration of 0.02 mg/mL. The interaction of TPPS with the two batches of MWCNTs has been investigated in citrate buffer solution (10 mM, pH 2.4) following two different mixing order procedures: (i) porphyrin-first protocol (PF) and (ii) porphyrin-last protocol (PL), consisting of the addition of a proper volume of MWCNTs dispersion to a diluted TPPS solution in citrate buffer and of the addition of TPPS from a stock solution to a diluted MWCNTs dispersion, respectively [43,53]. In some experiments MWCNT-EPA in citrate buffer has been previously mixed with NaCl (0.15 M), followed by the addition of TPPS. In all the experiments, the final concentration of TPPS was 5 µM.

UV-Vis spectra have been collected on a diode-array spectrophotometer Agilent model 8452. The circular dichroism (CD) spectra were recorded on a JASCO J-720 spectropolarimeter, equipped with a 450 W xenon lamp. CD spectra were corrected both for the cell and buffer contributions. A Jasco mod. FP-750 spectrofluorometer has been used to record fluorescence emission and Resonance Light Scattering (RLS) spectra. Emission spectra were not corrected for the absorption of the samples and a synchronous scan protocol with a right angle geometry was adopted for collecting RLS spectra [54]. All the aqueous dispersions were analysed by using a 1 cm optical path cuvette.

TGA was performed by using PerkinElmer Instruments Pyris1 TGA at a heating rate of 10 °C/min over the range from room temperature (r.t) to 1000 °C under N_2 atmosphere.

A TEM, JEM2100 LaB, working at 100 kV, and a digital Scanning transmission electron microscopy (STEM) set with BF & DF STEM Detectors plus SE/BSE detector (University of Exeter, UK) were used to investigate morphology of MWCNTs and MWCNTs/TPPS J-aggregates. Samples were prepared by evaporating ten drops of the aqueous dispersions of the investigated system more days after mixing (1–3 days) on 300 mesh holey-carbon coated copper grids.

3. Results and Discussion

Synthesis and Characterization of Amine Multiwalled Carbon Nanotubes

Amine multiwalled carbon nanotubes, MWCNT-EDA and MWCNT-EPA, were prepared by coupling of carboxylated MWCNTs (MWCNT-Ox) with tert-butyl 2-aminoethylcarbamate (EDABoc) or tetraethylenepentamine using EDC/HOBt in DMF according to Figure 1A,B, respectively. MWCNT-Ox were prepared by the oxidation (HNO_3/H_2SO_4 1:3, 6 h, 60 °C) of commercially available multiwalled carbon nanotubes according to a previously reported procedure [8,50].

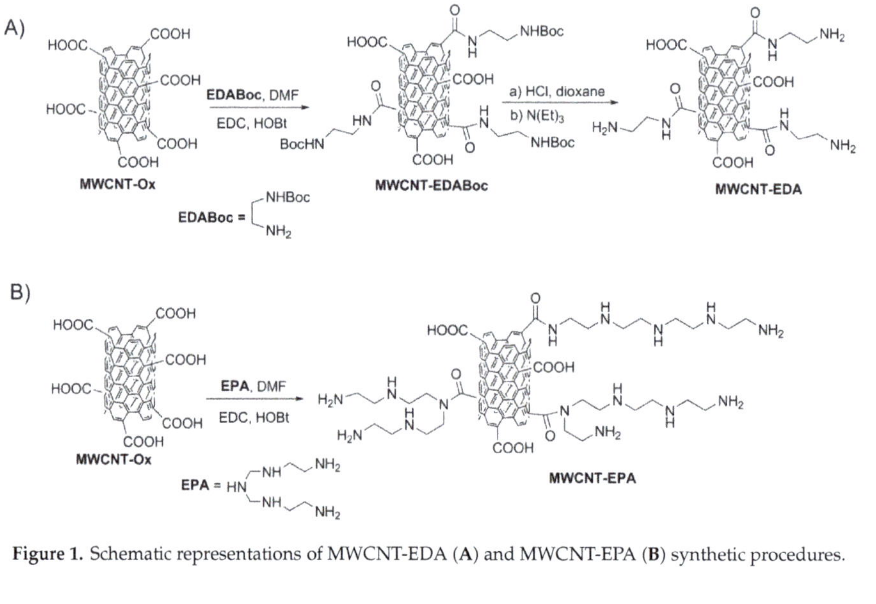

Figure 1. Schematic representations of MWCNT-EDA (**A**) and MWCNT-EPA (**B**) synthetic procedures.

The degree of functionalization of MWCNTs was investigated by TGA analysis (Figure 2) and the primary amines loadings was determined using the colorimetric Kaiser test [51,52].

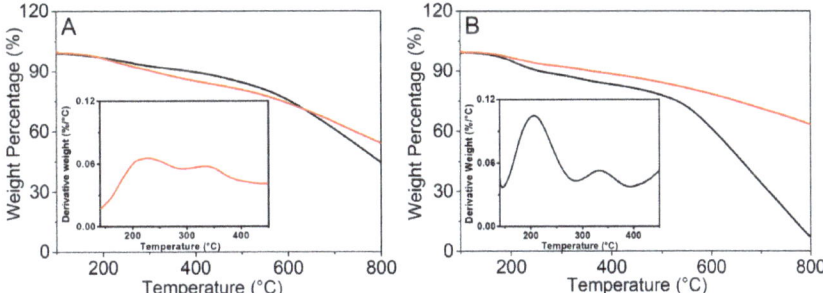

Figure 2. TGA profiles of MWCNT-Ox (dark line) and MWCNT-EPA (red line) (**A**). TGA profiles of MWCNT-EDABoc (dark line) and MWCNT-EDA (red line) (**B**). In the insets DTG curve of MWCNT-EPA (A) and MWCNT-EDABoc (B). TGA analyses were carried out under N_2 atmosphere.

The TGA curve of MWCNT-Ox shows a gradual weight loss of about 15.5% at 500 °C (Figure 2A). MWCNT-EPA TGA profile displays two weight loss steps in the range 100–400 °C, likely due to decomposition of polyamine alkyl chain (inset of Figure 2A). From TGA data, a weight loss of about 4.3% at 500 °C which roughly corresponds to 0.22 mmol/g of EPA (Figure 2A) has been estimated. By the correlation with the absorbance at 570 nm using the ninhydrin assay, an amount of free amino groups of 0.44 mmol/g has been determinated, probably suggesting a role of secondary amine groups in the amidation reactions (Figure 1B).

According to literature data, a higher thermal stability of MWCNTs containing free amine groups has been detected with respect to the MWCNTs sample containing Boc-amine groups (MWCNT-EDA vs. MWCNT-EDABoc, Figure 2B) [55,56]. The significant weight loss of MWCNT-EDABoc in the range 100–300 °C (see DTG, inset Figure 2B) can be attributed to the thermal decomposition and rearrangement of the tert-butoxyl groups. On the basis of TGA data, it was not realistic to determine the degree of functionalization in terms of weight loss ($\Delta m \approx 0.9$–1%) [57]. Thus, we have estimated the amount of free amine functional groups by the Kaiser test (0.22 mmol/g).

TEM analyses of functionalized MWCNTs indicated that the chemical functionalization with amine groups preserved the characteristic morphology of multiwalled tubes scaffold. An average external diameter of ~10 nm, corresponding to an average number of 8–10 layers were found. MWCNT-ox appeared strongly aggregates in bundles (Figure 3A), whereas well distinct isolated MWCNTs are observed in TEM image of amine functionalized MWCNTs (MWCNT-EDA, Figure 3B and MWCNT-EPA, Figure 3C). Moreover, all the functionalized MWCNTs (MWCNT-Ox, MWCNT-EPA and MWCNT-EDA) were shortened by oxidation: the length was reduced from the micrometre (pristine MWCNTs, see Supplementary Figure S1) to nanometre scale [8,50].

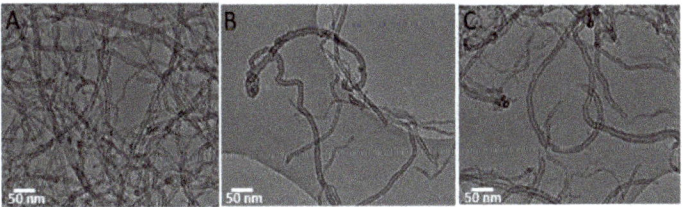

Figure 3. TEM images of MWCNT-Ox (**A**), MWCNT-EDA (**B**) and MWCNT-EPA (**C**).

In order to obtain MWCNT-EDA/TPPS J-aggregates hybrids, the two aforementioned mixing order protocols (both PF and PL) have been used under mild acidic condition. Whatever of the procedure employed, we found that the extinction features of free diacid porphyrin (B band centered at 434 nm and Q bands at 592 and 645 nm) remained unchanged even after 1 day (Figure 4A). No TPPS J aggregates were formed and no interaction between TPPS and MWCNT-EDA was revealed. This behavior could be ascribable to the presence of only one pendant amino group for chain in EDA, and this observation agrees with previous results on the role of the amine chain length in inducing the aggregation of TPPS [53].

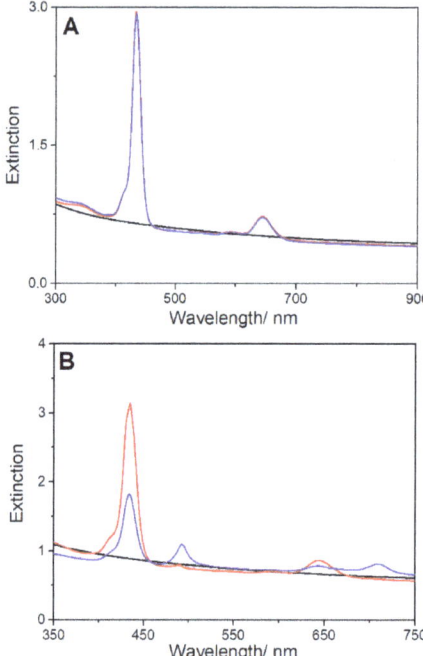

Figure 4. UV-Vis spectra of MWCNT-EDA (black line) and upon TPPS addition (red line), and 1 day after mixing (blue line) (**A**), and of MWCNT-EPA (black line), upon TPPS addition (red line), and 1 day after mixing (blue line) (**B**). Experimental conditions: [TPPS] = 5 µM; MWCNT-EDA or MWCNT-EPA = 0.02 mg/mL; 10 mM citrate buffer at pH 2.4; PL protocol; T = 298 K.

On the other hand, upon the addition of the porphyrin to the MWCNT-EPA dispersion, the UV-Vis profile shows the spectral signatures both of the diacid form of TPPS (B-band at 434 nm) and of J-aggregates evidenced by their typical extinction band arising at 489 nm. During the time, we observed the decrease of the intensity of the diacid band accompanied by the increase of the intensity of J-band, which furthermore undergoes a bathochromic shift from 489 to 493 nm. After one day, at the end of the aggregation process, UV-Vis spectrum of MWCNT-EPA/TPPS exhibits the B and Q bands ascribable to the residual monomeric diacid form of TPPS and J-aggregates (Figure 4B). It is noteworthy that the formation of J-aggregates does not occur under the same experimental conditions in absence of MWCNT-EPA, but it can be forced by decreasing the pH of the medium [42]. On the bases of the experimental evidences, we suggest that only EPA functionalized MWCNTs are able to trigger the TPPS aggregation process. This could be ascribable to the occurrence of an initial electrostatic interaction among a sufficient number of positively charged protonated amino groups on the CNTs surface and negatively charged sulfonated groups present in the periphery of the dyes [31,58]. Moreover, the observed red-shift of the extinction B-band of J-aggregates in the time could suggest a rearrangement of the aggregates due to an interaction with different amine-modified carbon nanotubes or their location in a different microenvironment with respect to the aqueous solution.

The emission spectrum (λ_{exc} 455 nm), at the end of the aggregation process, shows the typical fluorescence emission of the diacid form of TPPS centered at 669 nm. Further, upon excitation on the J-aggregates band (λ_{exc} 493 nm), an emission band at 717 nm can be also detectable (Figure 5). The different ratio of the bands intensity at 669 and 717 nm at the two distinctive excitation wavelengths is due to the fluorescence emission generated from the aggregated species. This evidence is confirmed by the related excitation spectra (Figure 5 inset) showing spectral features for both the diacidic TPPS and J-aggregates. However, the monomeric porphyrin is the predominant species in the excitation profile at both emission wavelengths, due to its longer lifetime value with respect to the J-aggregate one [59,60].

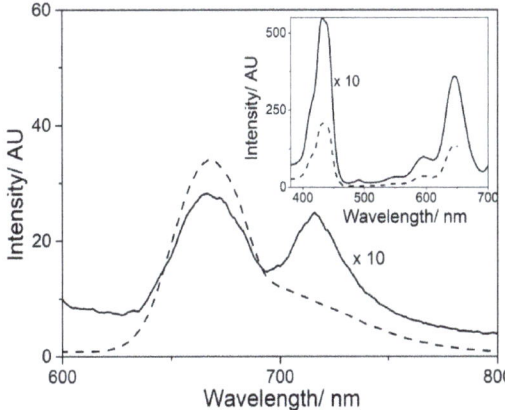

Figure 5. Fluorescence emission spectra of MWCNT-EPA/TPPS J-aggregates system 1 day after mixing (dashed and solid lines recorded at λ_{exc} = 455 and 493 nm, respectively) and, in the inset, the corresponding excitation spectra (dashed and solid lines recorded at λ_{em} = 669 and 717 nm, respectively). Experimental conditions: [TPPS] = 5 µM; MWCNT-EPA = 0.02 mg/mL; 10 mM citrate buffer at pH 2.4; PL protocol; T = 298 K.

The RLS spectra recorded soon after mixing shows a very sharp peak in the red region of the extinction band which increases in intensity at the end of the aggregation process. These findings agree with our previous results [42], pointing to the formation of self-assemblies of electronically coupled porphyrins, which cause a large enhancement of the resonant light scattering [54] at the red-edge of the extinction peak (Figure 6).

J-aggregates of TPPS induced by MWCNT-EPA were stable in acidic aqueous dispersion for a day or more after preparation. In difference with our previous findings on polyamine-mediated J-aggregates [30,31,61], the optical profiles are in line with the usual Frenkel exciton theory, rather than in terms of an extended network formed by the J-aggregates and amine modified MWCNTs in which dipole–dipole coupling among single porphyrins takes place.

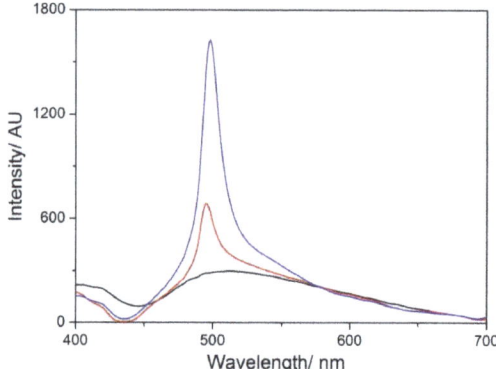

Figure 6. RLS spectra of MWCNT-EPA (black line), after TPPS addition (red line), and 1 day after mixing (blue line). Experimental conditions: [TPPS] = 5 µM; MWCNT-EPA = 0.02 mg/mL; 10 mM citrate buffer at pH 2.4; PL protocol; T = 298 K.

CD spectra were recorded after freshly mixing of the components and at the end of the aggregation process (Figure 7). As expected for achiral MWCNT-EPA, CD spectrum is silent before the addition of the chromophoric species, so confirming the absence of optical activity for amine modified CNTs. On the other hand, when TPPS was added a slight bisegnate positive Cotton effect in the aggregates absorption region has been observed. At the end of the aggregation process, an increase in intensity and a red shift of the CD profile were observed. This behavior, observed by means of spectroscopic and light scattering techniques, is due to the formation of large and rearranged structures as the result of the interactions among porphyrin aggregates and functionalized MWCNTs.

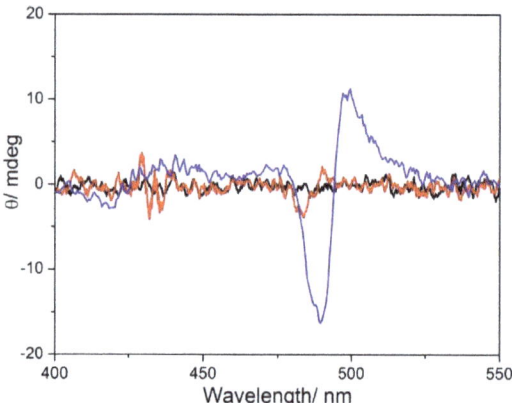

Figure 7. CD spectra of MWCNT-EPA (black line), soon after TPPS addition (red line), and 1 day after mixing (blue line). Experimental conditions: [TPPS] = 5 µM; MWCNT-EPA = 0.02 mg/mL; 10 mM citrate buffer at pH 2.4; PL protocol; T = 298 K.

In agreement with the spectroscopic characterization, representative TEM images of MWCNT-EPA/TPPS J-aggregates pointed to the coexistence of both MWCNT-EPA and TPPS J-aggregates in the same area (Figure 8 and Supplementary Figure S2). In particular, TPPS J-aggregates with an average diameter of 45 nm and length of 250–500 nm seem to be wrapped by separate nanotubes or bundles of MWCNT-EPA having an average external diameter of about 15 nm.

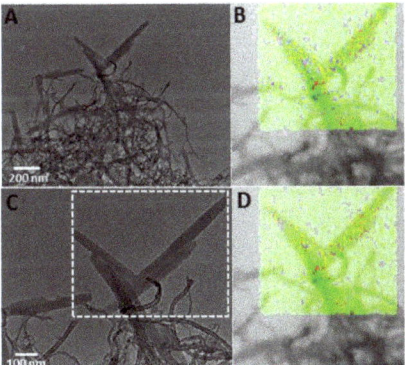

Figure 8. TEM images of MWCNT-EPA/TPPS J-aggregates at low (**A**) and high resolution (**C**): (**B,D**) corresponds to STEM analysis of total (C/N/O/S) and C/S merging of elements distribution respectively taken within the dashed region of the assemblies in (**C**) (for individuals colours of elements refers to (**B**) in Figure S3; see Materials and Methods for preparation conditions).

STEM analysis of MWCNT-EPA/TPPS J-aggregates shows the total elemental distribution pointing out the presence of carbon and oxygen for MWCNT-EPA, and carbon, oxygen, sulfur and nitrogen for TPPS J-aggregates (Supplementary Figure S3). Interestingly, in the marked area (Figure 8C), the total elements merging (Figure 8B) appears to be similar to the carbon/sulfur merging (Figure 8D). These results evidence the co-localization of TPPS J-aggregates and carbon nanotubes in the investigated samples. Since self-organization phenomenon is a hierarchical process, it is well known as the morphology of final aggregates can be controlled by the mixing order protocol [43]. In this framework, we performed experiments by adding amine carbon nanotubes to diacid TPPS (PF protocol). Surprisingly, at the end of the aggregation process, all the spectroscopic (Supplementary Figures S4 and S5) and morphological features (Supplementary Figure S6) show no change with respect to that observed by previous reagent mixing order protocol (PL). Because, in the case of the self-aggregation of neat TPPS in acidic conditions [43], the mixing order protocol is strictly related to the occurrence of porphyrin nucleation phenomena, here, we are prone to think that MWCNT-EPA could act as nucleation centers, inducing dye aggregation independently by the mixing order protocol.

In order to verify the stability of MWCNT-EPA/TPPS J-aggregates in mimicking physiological medium (NaCl 0.9% $w/w \cong$ 0.15 M), the system has been prepared by firstly dispersing MWCNT-EPA in NaCl 0.15 M aqueous solution and then adding TPPS. Under these conditions, the spectroscopic evidences of the final system remain almost unchanged (Figure 9) with respect to the unsalted solutions thus confirming the formation of chromophoric assemblies. The premixing of MWCNTs and NaCl, followed by the addition of porphyrin, seems to lead to a larger amount of J-aggregates due to the ionic strength effect [62]. Generally, optical stability for J-aggregates is difficult to achieve. Therefore, the use of surfactans or the entrapment in liposomes of the dye forming J-aggregates were experienced in literature [21]. In our case, MWCNTs induce, whatever the preparation procedure, the formation of stable J-aggregates able to retain their optical properties even after more days (1–3 days). Therefore, no further manipulation to preserve their pristine optical properties was necessary.

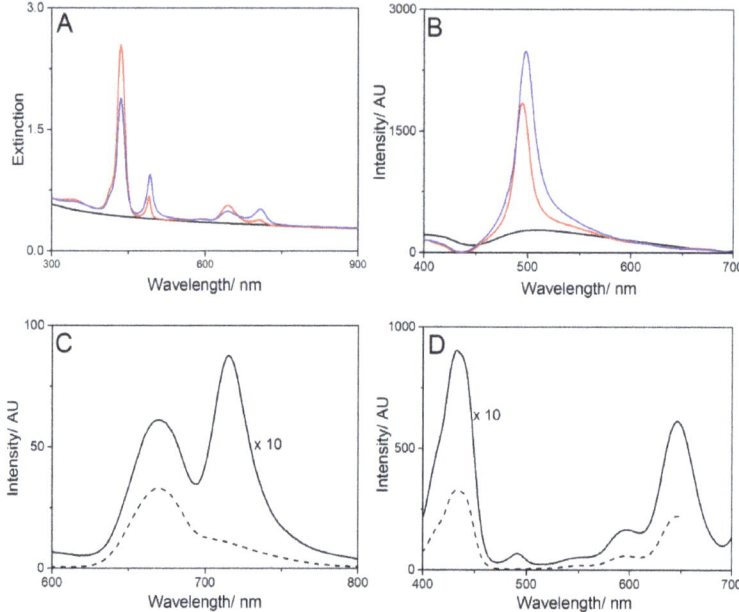

Figure 9. UV-Vis spectra (**A**) and RLS (**B**) of MWCNT-EPA (black line), upon TPPS addition (red line) and 4 h after mixing (blue line); (**C**) Fluorescence emission spectra of MWCNT-EPA/TPPS J-aggregates (dashed and solid lines at λ_{exc} = 455 and 492 nm, respectively) and (**D**) the corresponding excitation spectrum (dashed and solid lines at λ_{em} = 669 and and 717 nm, respectively). Experimental conditions: [TPPS] = 5 µM; MWCNT-EPA = 0.02 mg/mL; 10 mM citrate buffer at pH 2.4; [NaCl] = 0.15 M, PL protocol; T = 298 K.

Altogether, the combination of drug carrier ability of MWCNTs with the theranostic properties of porphyrins could allow the development of MWCNTs/TPPS J-aggregates nanohybrids for applications in biomedical field. Unlike from the others families of hybrid carbon nanomaterials [9], the potential applications in biological/pharmaceutical field of carbon based nanomaterials- porphyrins appear still scarcely investigated, especially for J-aggregates self-assemblies as well as their intracellular trafficking, therapeutic and imaging properties. In the literature, it was observed that hybrids nanomaterials based on porphyrinoids [24,63] were prepared at pH different from physiological conditions, and then treated with cells. In this context, future work will be devoted to studying the biocompatibility and cellular uptake of our hybrid MWCNTs/J-aggregates supramolecular systems. With these perspectives in mind, this research could lay the groundwork for the incoming biological assessment of MWCNT-EPA/TPPS J-aggregates.

4. Conclusions

MWCNTs can be easily functionalized by covalently introducing pendant amino-groups on their surface. In this paper we used ethylenediamine (EDA) and tetraethylenepentamine (EPA), which after coupling with carboxylic groups on the exterior walls, led to one and four protonable amino groups for chain, respectively. Under mild acidic conditions, the diacid form of TPPS is able to electrostatically bind to the surface and eventually aggregate. In line with our previous investigations on the ability of polyamines to trigger the aggregation of TPPS, the presence of only one pendant amino group (EDA) is not enough to induce the formation of TPPS J-aggregates, whatever the mixing protocol. On the other hand, when EPA functionality is present, these species are effective to generate stable MWCNT-EPA/TPPS J-aggregates nanohybrids and their general spectroscopic features are rather independent on the mixing protocol. A similar behavior was observed in solutions mimicking physiological medium (NaCl ≅ 0.15 M), whereby stable nanohybrids were also obtained. These systems exhibit remarkable optical features, and in this perspective could be considered for potential applications in phototherapy (by irradiating on extinction bands at 491 nm and/or at 709 nm) and/or bio-imaging (by exploiting the fluorescent emission band at 716 nm). In this respect, this class of amine-modified MWCNTs could be investigated as carriers of J-aggregates in biological environment. All these optical and structural properties make MWCNT-EPA/TPPS J-aggregates appealing for further considerations in theranostic.

Supplementary Materials: The following are available online at http://www.mdpi.com/2079-4991/10/4/669/s1, Figure S1: TEM image of pristine MWCNTs; Figure S2: TEM images of MWCNT-EPA/TPPS J-aggregates; Figure S3: STEM of MWCNT-EPA/TPPS J-aggregates; Figure S4: UV-Vis spectra of an aqueous solution of TPPS J-aggregates/MWCNT-EPA (PF protocol) and MWCNT-EPA TPPS J-aggregates (PL protocol); Figure S5: CD spectra of an aqueous solution of TPPS J-aggregates/MWCNT-EPA (PF protocol) and MWCNT-EPA/TPPS J-aggregates (PL protocol); Figure S6: TEM images of TPPS J-aggregates/MWCNT-EPA (PF protocol).

Author Contributions: Conceptualization, A.M., M.A.C., A.R. and L.M.S.; investigation, M.T., A.M., A.P. and A.C.; data curation, M.T., A.M. and A.P.; writing—original draft preparation, M.T., A.M. and L.M.S.; writing—review and editing, A.M., R.Z., M.A.C. and A.R.; visualization, R.Z., M.A.C. and A.R.; Supervision, L.M.S. All authors discussed the results and commented on the manuscript. All authors have read and agreed to the published version of the manuscript.

Funding: This research was funded by CNR (Project ISMN-CNR: Materials and Dispositives for Health and Life Quality) for financial support.

Acknowledgments: We are grateful to Yanqui Zhu and Hong Chang (University of Exeter) for kind assistance with TGA, TEM and STEM analyses.

Conflicts of Interest: The authors declare no conflict of interest.

References

1. Poudel, Y.R.; Li, W. Synthesis, properties, and applications of carbon nanotubes filled with foreign materials: A review. *Mater. Today Phys.* **2018**, *7*, 7–34. [CrossRef]
2. Marchesan, S.; Kostarelos, K.; Bianco, A.; Prato, M. The winding road for carbon nanotubes in nanomedicine. *Mater. Today* **2015**, *18*, 12–19. [CrossRef]
3. Negri, V.; Pacheco-Torres, J.; Calle, D.; López-Larrubia, P. Carbon Nanotubes in Biomedicine. *Top. Curr. Chem.* **2020**, *378*, 15. [CrossRef] [PubMed]
4. Costa, P.M.; Bourgognon, M.; Wang, J.T.W.; Al-Jamal, K.T. Functionalised carbon nanotubes: From intracellular uptake and cell-related toxicity to systemic brain delivery. *J. Control. Release* **2016**, *241*, 200–219. [CrossRef] [PubMed]
5. Aoki, K.; Saito, N. Biocompatibility and Carcinogenicity of Carbon Nanotubes as Biomaterials. *Nanomaterials* **2020**, *10*, 264. [CrossRef] [PubMed]
6. Battigelli, A.; Ménard-Moyon, C.; Da Ros, T.; Prato, M.; Bianco, A. Endowing carbon nanotubes with biological and biomedical properties by chemical modifications. *Adv. Drug Deliv. Rev.* **2013**, *65*, 1899–1920. [CrossRef]
7. Iannazzo, D.; Mazzaglia, A.; Scala, A.; Pistone, A.; Galvagno, S.; Lanza, M.; Riccucci, C.; Ingo, G.M.; Colao, I.; Sciortino, M.T.; et al. β-Cyclodextrin-grafted on multiwalled carbon nanotubes as versatile nanoplatform for entrapment of guanine-based drugs. *Colloids Surf. B Biointerfaces* **2014**, *123*, 264–270. [CrossRef]
8. Mazzaglia, A.; Scala, A.; Sortino, G.; Zagami, R.; Zhu, Y.; Sciortino, M.T.; Pennisi, R.; Pizzo, M.M.; Neri, G.; Grassi, G.; et al. Intracellular trafficking and therapeutic outcome of multiwalled carbon nanotubes modified with cyclodextrins and polyethylenimine. *Colloids Surf. B Biointerfaces* **2018**, *163*, 55–63. [CrossRef]
9. Piperno, A.; Scala, A.; Mazzaglia, A.; Neri, G.; Pennisi, R.; Sciortino, M.T.; Grassi, G. Cellular Signaling Pathways Activated by Functional Graphene Nanomaterials. *Int. J. Mol. Sci.* **2018**, *19*, 3365. [CrossRef]
10. Piperno, A.; Mazzaglia, A.; Scala, A.; Pennisi, R.; Zagami, R.; Neri, G.; Torcasio, S.M.; Rosmini, C.; Mineo, P.G.; Potara, M.; et al. Casting Light on Intracellular Tracking of a New Functional Graphene-Based MicroRNA Delivery System by FLIM and Raman Imaging. *ACS Appl. Mater. Interfaces* **2019**, *11*, 46101–46111. [CrossRef]

11. Vialla, F.; Delport, G.; Chassagneux, Y.; Roussignol, P.; Lauret, J.S.; Voisin, C. Diameter-selective non-covalent functionalization of carbon nanotubes with porphyrin monomers. *Nanoscale* **2016**, *8*, 2326–2332. [CrossRef] [PubMed]
12. Sprafke, J.K.; Stranks, S.D.; Warner, J.H.; Nicholas, R.J.; Anderson, H.L. Noncovalent Binding of Carbon Nanotubes by Porphyrin Oligomers. *Angew. Chem. Int. Ed.* **2011**, *50*, 2313–2316. [CrossRef] [PubMed]
13. Vizuete, M.; Gómez-Escalonilla, M.J.; Fierro, J.L.G.; Atienzar, P.; García, H.; Langa, F. Double-Wall Carbon Nanotube–Porphyrin Supramolecular Hybrid: Synthesis and Photophysical Studies. *ChemPhysChem* **2014**, *15*, 100–108. [CrossRef] [PubMed]
14. Magadur, G.; Lauret, J.-S.; Alain-Rizzo, V.; Voisin, C.; Roussignol, P.; Deleporte, E.; Delaire, J.A. Excitation Transfer in Functionalized Carbon Nanotubes. *ChemPhysChem* **2008**, *9*, 1250–1253. [CrossRef] [PubMed]
15. Zhang, H.; Bork, M.A.; Riedy, K.J.; McMillin, D.R.; Choi, J.H. Understanding Photophysical Interactions of Semiconducting Carbon Nanotubes with Porphyrin Chromophores. *J. Phys. Chem. C* **2014**, *118*, 11612–11619. [CrossRef]
16. Hasobe, T.; Fukuzumi, S.; Kamat, P.V. Ordered Assembly of Protonated Porphyrin Driven by Single-Wall Carbon Nanotubes. J- and H-Aggregates to Nanorods. *J. Am. Chem. Soc.* **2005**, *127*, 11884–11885. [CrossRef]
17. Aurisicchio, C.; Marega, R.; Corvaglia, V.; Mohanraj, J.; Delamare, R.; Vlad, D.A.; Kusko, C.; Dutu, C.A.; Minoia, A.; Deshayes, G.; et al. CNTs in Optoelectronic Devices: New Structural and Photophysical Insights on Porphyrin-DWCNTs Hybrid Materials. *Adv. Funct. Mater.* **2012**, *22*, 3209–3222. [CrossRef]
18. Devaramani, S.; Shinger, M.I.; Ma, X.; Yao, M.; Zhang, S.; Qin, D.; Lu, X. Porphyrin aggregates decorated MWCNT film for solar light harvesting: Influence of J- and H-aggregation on the charge recombination resistance, photocatalysis, and photoinduced charge transfer kinetics. *Phys. Chem. Chem. Phys.* **2017**, *19*, 18232–18242. [CrossRef]
19. Lutsyk, P.; Piryatinski, Y.; Shandura, M.; AlAraimi, M.; Tesa, M.; Arnaoutakis, G.E.; Melvin, A.A.; Kachkovsky, O.; Verbitsky, A.; Rozhin, A. Self-Assembly for Two Types of J-Aggregates: Cis-Isomers of Dye on the Carbon Nanotube Surface and Free Aggregates of Dye trans-Isomers. *J. Phys. Chem. C* **2019**, *123*, 19903–19911. [CrossRef]
20. Würthner, F.; Kaiser, T.E.; Saha-Möller, C.R. J-Aggregates: From Serendipitous Discovery to Supramolecular Engineering of Functional Dye Materials. *Angew. Chem. Int. Ed.* **2011**, *50*, 3376–3410. [CrossRef]
21. Song, X.; Zhang, R.; Liang, C.; Chen, Q.; Gong, H.; Liu, Z. Nano-assemblies of J-aggregates based on a NIR dye as a multifunctional drug carrier for combination cancer therapy. *Biomaterials* **2015**, *57*, 84–92. [CrossRef] [PubMed]
22. Miranda, D.; Huang, H.; Kang, H.; Zhan, Y.; Wang, D.; Zhou, Y.; Geng, J.; Kilian, H.I.; Stiles, W.; Razi, A.; et al. Highly-Soluble Cyanine J-aggregates Entrapped by Liposomes for In Vivo Optical Imaging around 930 nm. *Theranostics* **2019**, *9*, 381–390. [CrossRef] [PubMed]
23. Shakiba, M.; Ng, K.K.; Huynh, E.; Chan, H.; Charron, D.M.; Chen, J.; Muhanna, N.; Foster, F.S.; Wilson, B.C.; Zheng, G. Stable J-aggregation enabled dual photoacoustic and fluorescence nanoparticles for intraoperative cancer imaging. *Nanoscale* **2016**, *8*, 12618–12625. [CrossRef] [PubMed]
24. Liu, W.-J.; Zhang, D.; Li, L.-L.; Qiao, Z.-Y.; Zhang, J.-C.; Zhao, Y.-X.; Qi, G.-B.; Wan, D.; Pan, J.; Wang, H. In Situ Construction and Characterization of Chlorin-Based Supramolecular Aggregates in Tumor Cells. *ACS Appl. Mater. Interfaces* **2016**, *8*, 22875–22883. [CrossRef]
25. Cheng, M.H.Y.; Harmatys, K.M.; Charron, D.M.; Chen, J.; Zheng, G. Stable J-Aggregation of an aza-BODIPY-Lipid in a Liposome for Optical Cancer Imaging. *Angew. Chem. Int. Ed.* **2019**, *58*, 13394–13399. [CrossRef]
26. Gandini, S.C.M.; Gelamo, E.L.; Itri, R.; Tabak, M. Small angle X-ray scattering study of meso-tetrakis (4-sulfonatophenyl) porphyrin in aqueous solution: A self-aggregation model. *Biophys. J.* **2003**, *85*, 1259–1268. [CrossRef]
27. Collini, E.; Ferrante, C.; Bozio, R.; Lodi, A.; Ponterini, G. Large third-order nonlinear optical response of porphyrin J-aggregates oriented in self-assembled thin films. *J. Mater. Chem.* **2006**, *16*, 1573–1578. [CrossRef]
28. Collini, E.; Ferrante, C.; Bozio, R. Strong Enhancement of the Two-Photon Absorption of Tetrakis(4-sulfonatophenyl)porphyrin Diacid in Water upon Aggregation. *J. Phys. Chem. B* **2005**, *109*, 2–5. [CrossRef]
29. Castriciano, M.A.; Donato, M.G.; Villari, V.; Micali, N.; Romeo, A.; Scolaro, L.M. Surfactant-like behavior of short-chain alcohols in porphyrin aggregation. *J. Phys. Chem. B* **2009**, *113*, 11173–11178. [CrossRef]

30. Zagami, R.; Castriciano, M.A.; Romeo, A.; Trapani, M.; Pedicini, R.; Monsù Scolaro, L. Tuning supramolecular chirality in nano and mesoscopic porphyrin J-aggregates. *Dyes Pigments* **2017**, *142*, 255–261. [CrossRef]
31. Scolaro, L.M.; Romeo, A.; Castriciano, M.A.; Micali, N. Unusual optical properties of porphyrin fractal J-aggregates. *Chem. Commun.* **2005**, 3018–3020. [CrossRef]
32. Villari, V.; Fazio, B.; De Luca, G.; Trapani, M.; Romeo, A.; Scolaro, L.M.; Castriciano, M.A.; Mazzaglia, A.; Micali, N. Scattering enhancement in colloidal metal-organic composite aggregates. *Colloids Surf. A Physicochem. Eng. Asp.* **2012**, *413*, 13–16. [CrossRef]
33. Villari, V.; Fazio, B.; Micali, N.; De Luca, G.; Corsaro, C.; Romeo, A.; Scolaro, L.M.; Castriciano, M.A.; Mazzaglia, A. Light scattering enhancement in porphyrin nanocomposites. In *Proceedings of the International School of Physics Enrico Fermi*; Mallamace, F., Stanley, H., Eds.; IOS: Amsterdam, The Netherlands; SIF: Bologna, Italy, 2012; pp. 335–340. [CrossRef]
34. Castriciano, M.A.; Romeo, A.; Scolaro, L.M. Aggregation of meso-tetrakis(4-sulfonatophenyl)porphyrin on polyethyleneimine in aqueous solutions and on a glass surface. *J. Porphyr. Phthalocyanines* **2002**, *6*, 431–438. [CrossRef]
35. Micali, N.; Villari, V.; Romeo, A.; Castriciano, M.A.; Scolaro, L.M. Evidence of the early stage of porphyrin aggregation by enhanced Raman scattering and fluorescence spectroscopy. *Phys. Rev. E Stat. Nonlinear Soft Matter Phys.* **2007**, *76*. [CrossRef] [PubMed]
36. Castriciano, M.A.; Trapani, M.; Romeo, A.; Depalo, N.; Rizzi, F.; Fanizza, E.; Patanè, S.; Monsù Scolaro, L. Influence of Magnetic Micelles on Assembly and Deposition of Porphyrin J-Aggregates. *Nanomaterials* **2020**, *10*, 187. [CrossRef] [PubMed]
37. Trapani, M.; Castriciano, M.A.; Romeo, A.; De Luca, G.; Machado, N.; Howes, B.D.; Smulevich, G.; Scolaro, L.M. Nanohybrid Assemblies of Porphyrin and Au-10 Cluster Nanoparticles. *Nanomaterials* **2019**, *9*. [CrossRef] [PubMed]
38. Trapani, M.; De Luca, G.; Romeo, A.; Castriciano, M.A.; Scolaro, L.M. Spectroscopic investigation on porphyrins nano-assemblies onto gold nanorods. *Spectrochim. Acta Part A Mol. Biomol. Spectrosc.* **2017**, *173*, 343–349. [CrossRef]
39. Castriciano, M.A.; Leone, N.; Cardiano, P.; Manickam, S.; Scolaro, L.M.; Lo Schiavo, S. A new supramolecular polyhedral oligomeric silsesquioxanes (POSS)-porphyrin nanohybrid: Synthesis and spectroscopic characterization. *J. Mater. Chem. C* **2013**, *1*, 4746–4753. [CrossRef]
40. Villari, V.; Mazzaglia, A.; Trapani, M.; Castriciano, M.A.; De Luca, G.; Romeo, A.; Scolaro, L.M.; Micali, N. Optical enhancement and structural properties of a hybrid organic-inorganic ternary nanocomposite. *J. Phys. Chem. C* **2011**, *115*, 5435–5439. [CrossRef]
41. Hollingsworth, J.V.; Richard, A.J.; Vicente, M.G.H.; Russo, P.S. Characterization of the Self-Assembly of meso-Tetra(4-sulfonatophenyl)porphyrin (H2TPPS4−) in Aqueous Solutions. *Biomacromolecules* **2012**, *13*, 60–72. [CrossRef]
42. Castriciano, M.A.; Romeo, A.; Villari, V.; Micali, N.; Scolaro, L.M. Structural rearrangements in 5,10,15,20-tetrakis(4-sulfonatophenyl)porphyrin J-aggregates under strongly acidic conditions. *J. Phys. Chem. B* **2003**, *107*, 8765–8771. [CrossRef]
43. Romeo, A.; Castriciano, M.A.; Occhiuto, I.; Zagami, R.; Pasternack, R.F.; Scolaro, L.M. Kinetic control of chirality in porphyrin J-aggregates. *J. Am. Chem. Soc.* **2014**, *136*, 40–43. [CrossRef] [PubMed]
44. Occhiuto, I.G.; Zagami, R.; Trapani, M.; Bolzonello, L.; Romeo, A.; Castriciano, M.A.; Collini, E.; Monsù Scolaro, L. The role of counter-anions in the kinetics and chirality of porphyrin J-aggregates. *Chem. Commun.* **2016**, *52*, 11520–11523. [CrossRef] [PubMed]
45. Occhiuto, I.; De Luca, G.; Trapani, M.; Scolaro, L.M.; Pasternack, R.F. Peripheral Stepwise Degradation of a Porphyrin J-Aggregate. *Inorg. Chem.* **2012**, *51*, 10074–10076. [CrossRef]
46. Trapani, M.; Occhiuto, I.G.; Zagami, R.; De Luca, G.; Castriciano, M.A.; Romeo, A.; Scolaro, L.M.; Pasternack, R.F. Mechanism for Copper(II)-Mediated Disaggregation of a Porphyrin J-Aggregate. *ACS Omega* **2018**, *3*, 18843–18848. [CrossRef]
47. Liu, R.; Tang, J.; Xu, Y.; Zhou, Y.; Dai, Z. Nano-sized Indocyanine Green J-aggregate as a One-component Theranostic Agent. *Nanotheranostics* **2017**, *1*, 430–439. [CrossRef]
48. Qu, K.; Xu, H.; Zhao, C.; Ren, J.; Qu, X. Amine-linker length dependent electron transfer between porphyrins and covalent amino-modified single-walled carbon nanotubes. *RSC Adv.* **2011**, *1*, 632–639. [CrossRef]

49. Chen, J.; Collier, C.P. Noncovalent Functionalization of Single-Walled Carbon Nanotubes with Water-Soluble Porphyrins. *J. Phys. Chem. B* **2005**, *109*, 7605–7609. [CrossRef]
50. Cardiano, P.; Fazio, E.; Lazzara, G.; Manickam, S.; Milioto, S.; Neri, F.; Mineo, P.G.; Piperno, A.; Lo Schiavo, S. Highly untangled multiwalled carbon nanotube@polyhedral oligomeric silsesquioxane ionic hybrids: Synthesis, characterization and nonlinear optical properties. *Carbon* **2015**, *86*, 325–337. [CrossRef]
51. Kaiser, E.; Colescott, R.L.; Bossinger, C.D.; Cook, P.I. Color test for detection of free terminal amino groups in the solid-phase synthesis of peptides. *Anal. Biochem.* **1970**, *34*, 595–598. [CrossRef]
52. Ménard-Moyon, C.; Fabbro, C.; Prato, M.; Bianco, A. One-Pot Triple Functionalization of Carbon Nanotubes. *Chem. Eur. J.* **2011**, *17*, 3222–3227. [CrossRef] [PubMed]
53. Romeo, A.; Angela Castriciano, M.; Scolaro, L.M. Spectroscopic and kinetic investigations on porphyrin J-aggregates induced by polyamines. *J. Porphyr. Phthalocyanines* **2010**, *14*, 713–721. [CrossRef]
54. Pasternack, R.F.; Collings, P.J. Resonance Light-Scattering—A New Technique for Studying Chromophore Aggregation. *Science* **1995**, *269*, 935–939. [CrossRef] [PubMed]
55. Tuci, G.; Vinattieri, C.; Luconi, L.; Ceppatelli, M.; Cicchi, S.; Brandi, A.; Filippi, J.; Melucci, M.; Giambastiani, G. "Click" on Tubes: A Versatile Approach towards Multimodal Functionalization of SWCNTs. *Chem. Eur. J.* **2012**, *18*, 8454–8463. [CrossRef] [PubMed]
56. Iannazzo, D.; Piperno, A.; Ferlazzo, A.; Pistone, A.; Milone, C.; Lanza, M.; Cimino, F.; Speciale, A.; Trombetta, D.; Saija, A.; et al. Functionalization of multi-walled carbon nanotubes with coumarin derivatives and their biological evaluation. *Organ. Biomol. Chem.* **2012**, *10*, 1025–1031. [CrossRef]
57. Grassi, G.; Scala, A.; Piperno, A.; Iannazzo, D.; Lanza, M.; Milone, C.; Pistone, A.; Galvagno, S. A facile and ecofriendly functionalization of multiwalled carbon nanotubes by an old mesoionic compound. *Chem. Commun.* **2012**, *48*, 6836–6838. [CrossRef]
58. Castriciano, M.A.; Carbone, A.; Saccà, A.; Donato, M.G.; Micali, N.; Romeo, A.; De Luca, G.; Scolaro, L.M. Optical and sensing features of TPPS4 J-aggregates embedded in Nafion®membranes: Influence of casting solvents. *J. Mater. Chem.* **2010**, *20*, 2882–2886. [CrossRef]
59. Maiti, N.C.; Mazumdar, S.; Periasamy, N. J- and H-aggregates of porphyrins with surfactants: Fluorescence, stopped flow and electron microscopy studies. *J. Porphyr. Phthalocyanines* **1998**, *2*, 369–376. [CrossRef]
60. Maiti, N.C.; Ravikanth, M.; Mazumdar, S.; Periasamy, N. Fluorescence Dynamics of Noncovalently Linked Porphyrin Dimers, and Aggregates. *J. Phys. Chem. B* **1995**, *99*, 17192–17197. [CrossRef]
61. Micali, N.; Villari, V.; Scolaro, L.M.; Romeo, A.; Castriciano, M.A. Light scattering enhancement in an aqueous solution of spermine-induced fractal J-aggregate composite. *Phys. Rev. E Stat. Nonlinear Soft Matter Phys.* **2005**, *72*. [CrossRef]
62. Micali, N.; Villari, V.; Castriciano, M.A.; Romeo, A.; Scolaro, L.M. From fractal to nanorod porphyrin J-aggregates. concentration-induced tuning of the aggregate size. *J. Phys. Chem. B* **2006**, *110*, 8289–8295. [CrossRef] [PubMed]
63. Trapani, M.; Romeo, A.; Parisi, T.; Sciortino, M.T.; Patanè, S.; Villari, V.; Mazzaglia, A. Supramolecular hybrid assemblies based on gold nanoparticles, amphiphilic cyclodextrin and porphyrins with combined phototherapeutic action. *RSC Adv.* **2013**, *3*, 5607–5614. [CrossRef]

© 2020 by the authors. Licensee MDPI, Basel, Switzerland. This article is an open access article distributed under the terms and conditions of the Creative Commons Attribution (CC BY) license (http://creativecommons.org/licenses/by/4.0/).

Article

SERS Sensing Properties of New Graphene/Gold Nanocomposite

Giulia Neri [1], Enza Fazio [2,*], Placido Giuseppe Mineo [3,4], Angela Scala [1] and Anna Piperno [1,*]

[1] Department of Chemical, Biological, Pharmaceutical and Environmental Sciences, University of Messina, Viale F. Stagno D'Alcontres 31, I-98166 Messina, Italy
[2] Department of Mathematical and Computational Sciences, Physics Science and Earth Science, University of Messina, Viale F. Stagno D'Alcontres 31, I-98166 Messina, Italy
[3] Department of Chemical Sciences, University of Catania, V.le A. Doria 6, 95125 Catania, Italy
[4] Institute for Chemical and Physical Processes-National Research Council (IPCF-CNR), Viale F. Stagno d'Alcontres 37, I-98158 Messina, Italy
* Correspondence: enfazio@unime.it (E.F.); apiperno@unime.it (A.P.); Tel.: +39-090-6765173 (A.P.)

Received: 19 July 2019; Accepted: 28 August 2019; Published: 30 August 2019

Abstract: The development of graphene (G) substrates without damage on the sp^2 network allows to tune the interactions with plasmonic noble metal surfaces to finally enhance surface enhanced Raman spectroscopy (SERS) effect. Here, we describe a new graphene/gold nanocomposite obtained by loading gold nanoparticles (Au NPs), produced by pulsed laser ablation in liquids (PLAL), on a new nitrogen-doped graphene platform (G-NH$_2$). The graphene platform was synthesized by direct delamination and chemical functionalization of graphite flakes with 4-methyl-2-*p*-nitrophenyl oxazolone, followed by reduction of *p*-nitrophenyl groups. Finally, the G-NH$_2$/Au SERS platform was prepared by using the conventional aerography spraying technique. SERS properties of G-NH$_2$/Au were tested using Rhodamine 6G (Rh6G) and Dopamine (DA) as molecular probes. Raman features of Rh6G and DA are still detectable for concentration values down to 1×10^{-5} M and 1×10^{-6} M respectively.

Keywords: graphene/gold nanocomposite; SERS; Dopamine; Rhodamine 6G

1. Introduction

Since their discovery, graphene materials (G), due to their outstanding physicochemical properties [1], have generated huge interest in numerous fields including biomedicine, electronics, sensing, energy, etc. [2–8]. They have been proposed as drug delivery systems for photothermal [9] and photodynamic therapy [10], as scaffold in tissue engineering [11], and as materials for biosensing [12,13]. Recently, G and its functionalized derivatives have been investigated as substrates for SERS (surface enhanced Raman spectroscopy) applications [14,15], a versatile technique that enables the rapid detection of various types of molecules [16,17].

Metal nanoparticles (i.e., Cu, Ag, gold nanoparticles (Au NPs)) are the most extensively studied SERS-active substrates since their collective electronic excitations, namely surface plasmons, are very interesting for a large variety of applications. Localized surface plasmon resonance excitation in Ag and Au NPs produces strong extinction and scattering spectra, resulting in amplification of the electric field (E) near the particle surfaces such that $|E|^2$ can be 100–10,000 times greater in intensity than the incident field, which acts on a spatial range of 10–50 nm. These effects are mainly influenced by two factors: (i) NPs morphology (in terms of size and shape) and (ii) local dielectric environment [18–20].

Two main mechanisms are involved in Raman signal enhancement: The electromagnetic mechanism (EM), due to the strong amplification of the local EM field [21], and the chemical effect (CM) that involves the creation of new electronic states generated by the interaction between the metal and the molecules adsorbed on it [22]. Such new electronic states allow for resonant Raman scattering processes; the control of the distances among the localized surface plasmons, on a sub-nanometer scale, is a critical parameter to control the inter-particle optical coupling and therefore, the efficiency of SERS response [23].

Recently, engineered G have attracted a huge amount of attention as platforms for biological SERS sensing [24]. G offer a large flat surface to adsorb molecules through π–π interactions determining the manner in which molecules bind with the surface which, in turn, determines the symmetry of the molecules and the effective charge transfer [25]. However, G alone provides a limited enhancement factor [26] while the combination of specifically designed G with metallic NPs is an interesting strategy to obtain new materials with synergistic effect and improved SERS sensing performance. The high compatibility of G with metal noble NPs is mainly due to: (i) Transparency to laser light and localized plasmonic fields; (ii) high thermal conductivity; and (iii) appropriate dielectric strength that confines the plasmonic field [27,28].

Despite the potentiality of these hybrid systems, the critical point of G-based SERS substrates regards the development of chemical strategies avoiding the damage of the G sp^2 network, keeping an electron high mobility and, at the same time, enabling the tuning of the interactions with plasmonic surface to finally enhance SERS effects. Generally, 2D materials were obtained by liquid chemical exfoliation of related 3D stratified bulk materials, processes that required the presence of intercalation agents and ultrasonication treatment. To overcome the long processing times and to guarantee the quality of 2D substrates, the exfoliation methods have been continuously implemented [29,30]. Recently, we have developed a straightforward method for the direct delamination of graphite flakes into functionalized G with preserved sp^2 network [31]. G-MNPO platform (Figure 1) [31], obtained by solvent-free 1,3-dipolar cycloaddition reaction of 4-methyl-2-p-nitrophenyl oxazolone with graphite, was selected for the development of a new nitrogen-doped graphene network (G-NH$_2$). The amine groups, obtained by reduction of p-nitrophenyl group on the Δ-1-pyrrolidine rings, were envisaged as anchoring sites for Au NPs. Here, we report the synthesis and characterization of graphene/gold nanocomposite (G-NH$_2$/Au) obtained by mixing G-NH$_2$ and Au NPs. Au NPs were produced by pulsed laser ablation in liquids (PLAL) technique that allowed the production of metal NPs in a variety of solvents with tuned size and optical properties [32,33]. No surfactant is needed to stabilize the colloids obtained by PLAL, and the NPs are extremely pure without any post-synthesis treatment [31]. To the best of our knowledge, no data have been reported in the literature about the SERS properties of G/Au platforms, where Au NPs were produced by PLAL technique.

The chemical composition and the morphology of G-NH$_2$ and G-NH$_2$/Au platforms were investigated by micro-Raman and X-ray photoelectron (XPS) spectroscopies, scanning transmission electron microscopy (STEM), and thermogravimetric analysis (TGA).

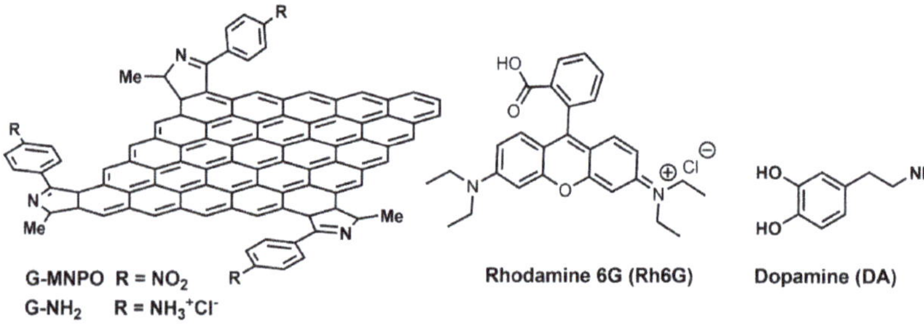

Figure 1. Schematic representation of G-MNPO and G-NH$_2$. Chemical structure of Rhodamine 6G (Rh6G) and Dopamine (DA).

The G-NH$_2$/Au dispersion was transferred onto a glass slide to obtain a uniform nanostructure thick film and its SERS properties were tested using Rhodamine 6G (Rh6G) and Dopamine (DA) as molecular probes (Figure 1).

2. Materials and Methods

2.1. Materials

Graphite flakes, Dopamine, Rhodamine 6G, solvents, and other reagents were purchased from Sigma Aldrich, (Milan, Italy); gold target (high purity, 99.99%) was purchased from Mateck srl (Jülich, Germany).

2.2. Synthesis of G-NH$_2$

G-MNPO was prepared according to the synthetic method already reported [31]. Further, 240 mg of G-MNPO, (0.09 mmol of NO$_2$) were homogenously dispersed in 30 mL H$_2$O by sonication (30 min). NaBH$_4$ (100 mg, 2.63 mmol) was added and the reaction mixture was stirred at 80 °C for 12 h. Afterwards, the reaction mixture was cooled to room temperature (r.t.), acidified to pH 3 by addition of a HCl 1M solution, and stirred for 1 h at r.t. G-NH$_2$ was recovered by filtration under vacuum (Millipore 0.1 µm) and it was purified by washing with 1:1 water/ethanol mixture. Finally, the residue was dried at ~60 °C to recover 185 mg of G-NH$_2$.

2.3. Synthesis of Au NPs by PLAL

Au water colloids were prepared according to previously reported procedure [34] using the 532 nm second harmonic emission wavelength of a Nd:YAG laser (Tempest- Laser Point srl, Milan Italy) operating at a repetition rate of 10 Hz (pulse length: 5 ns).

2.4. Synthesis of G-NH$_2$/Au

First, 20 mL of Au NPs were added to a dispersion of G-NH$_2$ (52 mg) in water (2 mL), obtained by sonication for 10 min, and the mixture was ultrasonicated (65% W) for 30 min. The reaction mixture was filtered at reduced pressure (Millipore 0.1 µm), the solid was repeatedly washed with water, and after drying at ~60 °C, 44 mg of G-NH$_2$/Au were recovered.

2.5. Preparation of G-NH$_2$/Au SERS Platform

The aqueous dispersion of G-NH$_2$/Au (5 mg/mL) was deposited onto a glass slide using the conventional aerography spraying technique. The aerography spraying system is made up by a high-pressure air brush with interchangeable nozzles of different sizes. During the deposition, the nozzle is continuously moved to ensure a uniform distribution on the substrate. The spraying is carried out in a deposition chamber equipped with a heated substrate holder and an excess vapors removal system to guarantee standard and reproducible conditions. The GNH$_2$/Au SERS platform was tested for Rhodamine 6G (Rh6G) at concentrations of 1×10^{-3}, 2×10^{-4}, 5×10^{-5} M and for Dopamine (DA) at concentrations of 1×10^{-3}, 2×10^{-4}, 5×10^{-5}, and 5×10^{-6} M. Rh6G and DA solutions were prepared using deionized water. The excitation sources were the 532 nm and 638 nm diode laser lines. The substrates were dipped in these solutions for 30 min and then taken out for free drying, after which the surface enhanced Raman (SERS) spectra were collected.

2.6. Samples Characterization

Thermal gravimetric analysis (TGA) profiles were acquired Perkin-Elmer Pyris TGA7 in the temperature range of 50–1000 °C. G-NH$_2$ or G-NH$_2$/Au (about 5 mg) were placed in a platinum pan and kept at 25 °C under a 60 mL min^{-1} air flow until balance stabilization (balance sensitivity was 0.01 mg), and subsequently heated with a scan rate of 10 °C min^{-1} under the same air flux. The calibration of instrument was settled according to previously reported procedure [31].

X-ray photoelectron spectroscopy was used to determine the surface elemental composition of the material and their bonding configurations. The spectra were acquired using a K-Alpha system (Thermo-Scientific, Germany) equipped with a monochromatic Al-Kα source (1486.6 eV), and operating in constant analyzer energy mode (pass energy: 200 eV), according to previously reported protocol [35]. Samples (G-NH$_2$, G-NH$_2$/Au, AuNPs) were deposited on a nickel grid to carry out scanning transmission electron microscopy (STEM) using a ZEISS instrument Merlin-Gemini 2 column (Merlin-Gemini, Germany), operating at primary voltage of 30 kV and at the working distance of 4 mm.

Raman spectra were acquired using the Horiba XploRA spectrometer (HORIBA Instruments, Milan, Italy) coupled with an optical microscope equipped with the 50X and 100X objectives. The excitation wavelengths used were 532 nm and 638 nm coming from solid diode lasers. The integration time was varied from 5 to 120 s, with an accumulation time of 2s, in order to optimize the signal to noise.

UV-vis optical absorption spectra of the Au and G-NH$_2$/Au samples were recorded using quartz cells and a Perkin Elmer (Lambda 750 model) spectrometer (Perkin Elmer, Milan, Italy) working in the 300–900 nm range.

3. Results

3.1. G-NH$_2$/Au SERS Platform

SERS platform based on graphene/gold nanocomposite (G-NH$_2$/Au) was obtained through a procedure involving: (i) Synthesis of G-NH$_2$ by reduction of G-MNPO; (ii) Preparation of G-NH$_2$/Au and (iii) Deposition of G-NH$_2$/Au onto a glass slide by an aerography spraying probe (Figure 2).

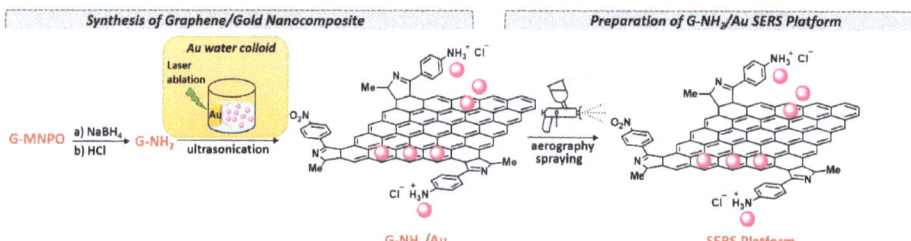

Figure 2. Preparation of G-NH$_2$/Au surface enhanced Raman spectroscopy (SERS) platform.

G-MNPO was prepared according to the synthetic method previously reported [31], carrying out the cycloaddition reaction at the molar ratio of 1:7 flake graphite/oxazolone. The experimental conditions for G-MNPO synthesis were optimized to obtain a substrate with a large surface area and a high degree of functionalization (0.037 mmol of NO$_2$/100 mg). G-MNPO was reduced with NaBH$_4$ and converted in the protonated salt by treatment with hydrochloric acid. The cationic centers on G-NH$_2$ surfaces increased the G dispersibility in water and guaranteed a better interaction with Au NPs. G-NH$_2$/Au nanocomposite was obtained by mixing, under ultrasonication treatment, the aqueous dispersion of G-NH$_2$ with the freshly prepared colloidal dispersion of Au NPs [34]. Finally, the aqueous dispersion of G-NH$_2$/Au was deposited onto the glass slide.

3.2. Characterization of Graphene/Gold Nanocomposite (G-NH$_2$/Au)

The content of Au NPs on G was estimated by TGA under air atmosphere (Figure 3). TGA profiles of G-NH$_2$ and G-NH$_2$/Au showed a high thermal stability without significant weight loss under 600 °C, indicating the absence of labile oxygen-containing functional groups. TGA profile of G-NH$_2$ showed a decomposition between 750 °C and 900 °C, with a complete decomposition of carbon at temperatures higher than 900 °C; whereas the G-NH$_2$/Au profile showed a lower decomposition temperature between 600 °C and 800 °C and the decomposition of carbon content became remarkable at 800 °C. The lower thermal decomposition of G-NH$_2$/Au compared to G-NH$_2$ could be attributed to the presence of Au NPs that increased the interlayer spacing and porosity of G-NH$_2$/Au. The residual mass of 7.29% indicated the loading of Au NPs on G-NH$_2$/Au nanocomposite.

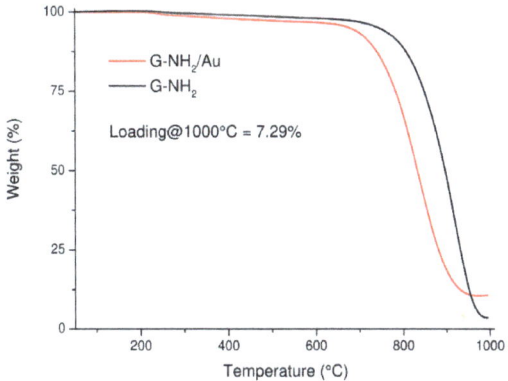

Figure 3. TGA profiles of G-NH$_2$ and G-NH$_2$/Au under air atmosphere.

Detailed information about the functionalities on G surfaces were obtained by XPS analysis. The wide scan spectra of G-NH$_2$ and G-NH$_2$/Au were reported in Figure 4a, with the Au 4f profile in the inset. This profile was characterized by well-separated spin-orbit components (Δ = 3.7 eV) where the Au 4f peak was centered at the binding energy of 84.0 eV, which is characteristic of the metal Au species. The Au, C, O, and N relative atomic percentages are reported in Table 1. The Au weight content percentage calculated by XPS was in good agreement with TGA data (Table 1). The N 1s high-resolution profile of G-NH$_2$ (Figure 4b) showed the presence of two peaks centered at about 400 eV, attributed to N=C and –NH$_3^+$ species, and at 407 eV due to NO$_2$. The lower contribution of the peak at 407 eV in G-NH$_2$ sample, compared with G-MNPO (20.05% vs. 41.18%, see Figure 4b and Table 1), indicated a good reduction of nitro groups into amino groups. The decrease of the oxygen content after the reduction reaction (19% vs. 7.4%, see Table 1) was connected with the changes observed by N 1s profile.

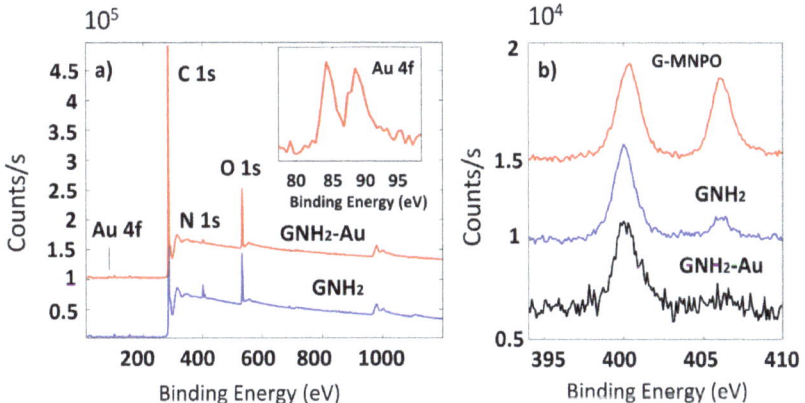

Figure 4. (**a**): XPS wide scan of G-NH$_2$ and G-NH$_2$/Au samples and line shapes of Au 4f (inset). (**b**): N 1s photoelectron deconvoluted line shapes of G-MNPO, G-NH$_2$, and G-NH$_2$/Au.

Table 1. Atomic content percentage for G-MNPO, G-NH$_2$, and G-NH$_2$/Au samples as determined by XPS analysis and N 1s percentage determined by deconvolution of XPS N 1s band. Weight content percentage of G-NH$_2$/Au calculated by XPS values (at the bottom).

Atomic Content Percentage Determined by XPS Analysis					N 1s Content Percentage Determined by Deconvolution of XPS N 1s Band	
Sample	Au	C	O	N	N 1s (N=C, NH$_3^+$)	N 1s (NO$_2$)
G-MNPO	-	74.8	19.0	6.2	58.82	41.18
G-NH$_2$	-	89.3	7.4	3.3	79.95	20.05
G-NH$_2$/Au	0.5	89.0	9.4	1.1	83.66	16.34
Weight content percentage calculated by XPS values						
G-NH$_2$/Au	7.5	80.2	11.3	1.2		

C 1s profiles of G-MNPO, G-NH$_2$, and G-NH$_2$/Au were deconvolved considering six spectral components: A main contribution at 284.5 eV attributed to C=C/C–C in the aromatic ring, and four other contributions, at higher binding energies, corresponding to carbon atoms bonded to nitrogen (C–N) and oxygen (C–OH, C–O, C=O) centered at 285.2, 286.3, 288.7, and 288.9 eV, respectively. The contribution at about 291.0 eV referred to π–π* bonds (Figure 5).

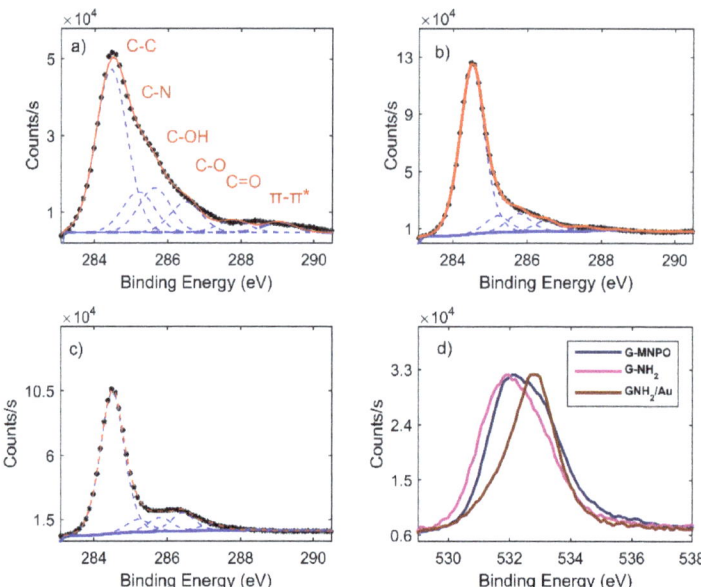

Figure 5. C 1s photoelectron deconvoluted line shapes: (**a**) G-MNPO, (**b**) G-NH$_2$, (**c**) G-NH$_2$/Au. (**d**) O 1s photoelectron deconvoluted line shapes of G-MNPO (blue), G-NH$_2$ (violet), G-NH$_2$/Au (brown).

Morphological information about the size and distribution of Au NPs within the G layers was obtained by electron microscopy analyses. STEM images (Figure 6) showed homogeneously distributed exfoliated G layers and various dimensional transparent sheets, in several portions of the sample, stacked onto each other, with a thickness of about 2–3 nm. Moreover, Au NPs characterized by an average size of 15 nm were mainly distributed at the edges of the G layers.

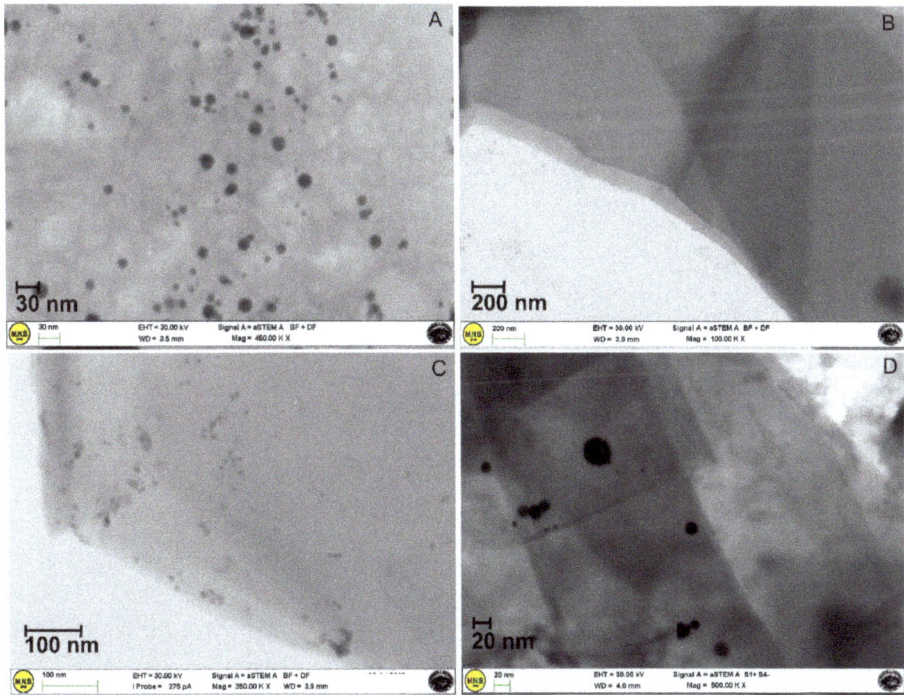

Figure 6. STEM images. (**A**) Au NPs produced by the green pulsed laser ablation technique in water at the laser fluence F of 1.5 J/cm^2 and the irradiation time t of 20 min. The NPs are nearly spherical in shape with a mean diameter of 15 nm; (**B**) G-NH$_2$ network with exfoliated G layers and transparent sheets, stacked onto each other, with a thickness of about 2–3 nm; (**C,D**) G-NH$_2$/Au platform, with Au NPs embedded within the overlapped thin layers of graphene.

In order to investigate the SERS enhancement of G-NH$_2$/Au, Raman spectroscopy analysis was performed (Figure 7). The Raman spectrum of G-NH$_2$ showed the G and 2D feature bands at 1580 and 2720 cm^{-1}, respectively (Figure 7). The very weak D-peak was indicative of the high quality of G and the 2D band splitting indicated the presence of a multilayers system. All these Raman contributions were also evident in the G-NH$_2$/Au nanocomposite, however some relevant differences were detected (Figure 7). Firstly, the increase of the intensity of all the peaks was observed, including a D band centered at about 1350 cm^{-1}, as a result of a certain degree of disorder induced by Au NPs insertion within G layers. The strong electric field gradient induced by the metallic NPs determined an overall change of the dipole moment during the vibration, even in the absence of a polarizability change. On the other hand, when G and Au NPs were in close proximity, some Raman forbidden peak appeared, namely the D' and the D+G contributions at about 1616 cm^{-1} and 2925 cm^{-1}, respectively. These evidences can be determined by: (i) The insertion of Au NPs on G-NH$_2$ platform, mainly at the edges of G layers (functionalized area of G layers) as suggested by computational studies [31] and (ii) reduced size of layers due to the mechanical effect of ultrasonication treatment adopted for the preparation of the nanocomposite. Moreover, the decrease of I_G/I_{2D} ratio, from 1.58 to 0.98, pointed out a better exfoliation of G-NH$_2$/Au with respect to G-NH$_2$; the insertion of Au NPs between G-NH$_2$ layers probably promoted their separation. Finally, the shifting of G and 2D bands suggested the anchorage of Au NPs on G surface. Raman signal was collected at several different sample locations to take into account the Au spatial homogeneity distribution within the nanoplatform. No significant changes were observed from one point to another one, which indicated that Au NPs were almost uniformly distributed within and/or on G layers.

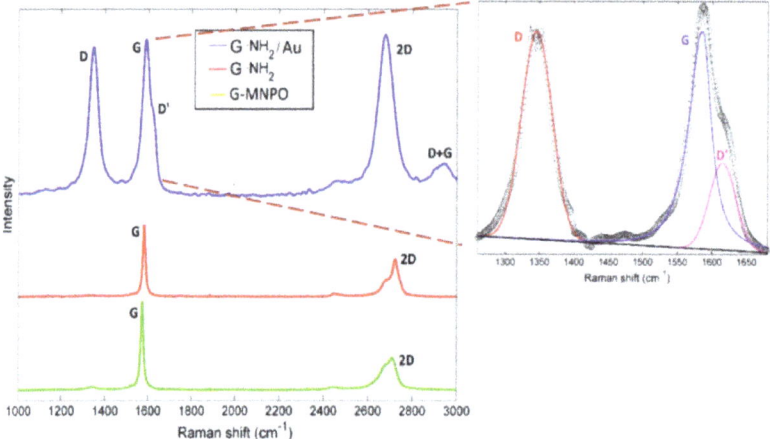

Figure 7. Raman spectra of G-MNPO, G-NH$_2$, G-NH$_2$/Au; the deconvolution of the D and G bands of the G-NH$_2$/Au is reported in the inset.

3.3. SERS Properties of G-NH$_2$/Au Platform

In order to test the SERS properties, the platforms (G-NH$_2$ and G-NH$_2$/Au) were immersed for 30 min in Rh6G aqueous solutions at different concentrations (1×10^{-3}, 2×10^{-4}, 5×10^{-5} M) and then air dried. Raman spectra were acquired using two different excitation diode laser lines (532 nm and 638 nm). UV-vis spectroscopy was exploited to determine the appropriate laser wavelength for resonant excitation of the localized surface plasmon. In fact, SERS is more effective when incident radiation falling on the nanostructured substrate is completely absorbed by metal NPs, so that excitation of the localized surface plasmon can take place. The field enhancement is greatest when the plasmon frequency is in resonance with the incident radiation. In Figure 8b, the optical absorption spectra of the freshly prepared Au NPs and of G-NH$_2$/Au were reported. Au NPs in water showed the characteristic Au surface plasmon resonance (SPR) band at 522 nm, due to the coherent oscillations of surface electrons interacting with an external electromagnetic field; whereas a red-shift (from 522 to 548 nm) and a decrease of the SPR intensity was observed in the G-NH$_2$/Au sample, suggesting an increase of the spatial distance between each Au NPs and the others, due to their dispersion into each G foil and/or within the G layers. Moreover, a charge transfer from Au NPs to G occurred, resulting in a decrease in electron density, which, in turn, contributed to the red-shift and the intensity decrease of the SPR band. Moreover, it is well known that the coating of gold surface with graphene modifies the propagation constant of surface plasmon polariton (SPP), thereby changing the sensitivity to refractive index change and, in turn, the optical response of the entire system [36].

SERS spectra, acquired using the 638 nm laser excitation, showed the well-defined Raman Rh6G peaks at about 615, 777, 1189, 1314, 1366, 1513, and 1651 cm^{-1} (Figure 8a). The feature at 615 cm^{-1} was assigned to the C–C–C in-plane bending mode, the peak at 777 cm^{-1} to the C–H out-of-plane bending mode and the residual peaks to the aromatic stretching vibrations of C atoms. Raman features were clearly observable at 10^{-3} M concentration and less evident, but still visible, for lower Rh6G concentration values (down to 1×10^{-5} M). On the other hand, if the Rh6G aqueous solution was deposited onto a G-NH$_2$ bare platform (i.e., without Au NPs), no Raman activity was detected even at a 10^{-3} M concentration. The Raman spectrum of G-NH$_2$ was characterized only by broad asymmetric and low-intensity bands, at around 1580 cm^{-1} (referred as G band) and near 1330 cm^{-1} (referred as D band), typical of carbon-based materials. The Rh6G SERS spectra obtained on the G-NH$_2$/Au platform were very similar to that obtained by using a substrate made from Au nanostructured film (Figure 8b). As a final remark, we observed that, by using a 532 nm laser excitation, no Raman signals could be collected in all the tested conditions. This unusual behavior can be explained taking into account the red-shifted observed SPR optical absorption, which certainly reduced the SERS effect. Summarizing, G layers positively influenced both the EM and the CM coupling, enhancing the SERS process due to the interesting optical properties, nanostructures high surface/volume ratio, and a great affinity between G and Au NPs.

The ability of the G-NH$_2$/Au nanocomposite to detect the biomolecules was tested using DA in a label-free configuration (Figure 9). DA is adsorbed on G surface through π–π stacking interactions [37]. The high surface area of G supporting the DA adsorption and diffusion processes was the primary condition for the efficient sensing of DA by SERS.

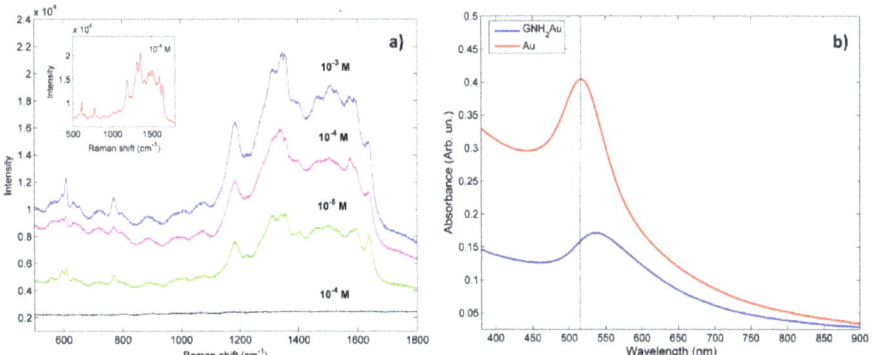

Figure 8. (**a**) SERS spectra of Rh6G (10^{-5}, 10^{-4} and 10^{-3} M) onto G-NH$_2$/Au platform; G-NH$_2$ platform (10^{-4} M, black line) and AuNPs film (10^{-4} M Rh6G reported in the inset), by using the 638 nm laser excitation. (**b**) Optical absorption spectra of freshly prepared Au NPs (red line) and G-NH$_2$/Au (blue line, 0.4 mg/mL).

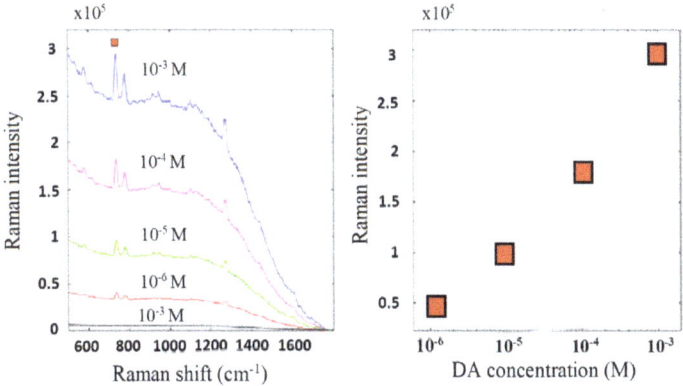

Figure 9. SERS spectra of DA tested at different concentrations (10^{-6}, 10^{-5}, 10^{-4}, and 10^{-3} M) onto the G-NH$_2$/Au platform along with the control test onto the G-NH$_2$ platform (10^{-3} M, black line), on the left. The Raman intensity signal trend onto DA concentration is reported on the right. The excitation is the 638 nm laser line.

SERS spectra of DA showed several characteristic peaks centered at about 608, 767, 1349 cm^{-1}. It is worth noting that other weak Raman features in the 1050–1300 cm^{-1} and 1500–1800 cm^{-1} regions can be detected despite the remarkable Raman background. The observed peaks were ascribed to the outside surface deformation of breathing, bending, and stretching vibrations of CH ring, bending vibration of NH, and aromatic C=C, respectively [38]. It is plausible that both EM and CM were involved in SERS signals, as already observed in reduced graphene oxide/silver nano-triangle sol substrate [39,40].

Au NPs between G layers behaved as "hot spots", which allowed the detection of DA Raman signals, not observed in the G-NH$_2$ bare platform. Since the distribution of Au NPs within G sheets played a fundamental role in determining SERS response and that it is known from the literature that one of the problems that still remain in question is the reproducibility and the repeatability of the spectra at low concentrations, we acquired SERS spectra in different points of the SERS substrate and at different times. We observed, point to point, very minimal variations in the intensity of some DA characteristic peaks without the degradation of nanocomposites.

4. Discussion

The results of this work has demonstrated the great potentiality for SERS applications of functionalized G obtained by covalent modification [41]. The cycloaddition protocol (i.e., 1,3-dipolar cycloaddition between mesoionic compounds and graphite) [31] furnished a G network decorated with Δ-1-pyrrolidine rings, mainly in the edge defected sites. This approach incorporated several interesting advantages: (i) Avoided the damage of sp^2 G network; (ii) provided functionalized G with high degree of functionalization (i.e., 4.6%) and the NO$_2$ group on pyrrolidine rings was reduced in good yield (XPS data indicated the reduction of almost half of the nitro groups in NH$_2$ groups); (iii) the amine groups assisted the anchorage of Au NPs produced by PLAL on G surface; (iv) the dispersibility in water of the G-NH$_2$ nanocomposite was enough for its deposition onto glass slide by aerography spraying technique. Au NPs included on G resulted in a hybrid nanocomposite (G-NH$_2$/Au) that combined the stronger plasmonic-based EM of Au with the superior stability, adsorption, and quenching of G. This nanocomposite revealed strong interactions between the two entities. From its spectral features, the origin of these interactions could be attributed primarily to the strong electric field gradient induced by Au NPs that determined an overall change of the dipole moment during the vibration, even in the absence of a polarizability change. The SERS activity of the assembled Au NPs with the graphene platform is justified in terms of "hot spots". Au NPs within the G-NH$_2$ structure are "confined" to a certain region sensitive to the Raman scattering [42,43]; localization of light as surface plasmons in noble metal nanostructures enables their potential role in antennas, single molecule detection, and surface-enhanced Raman. Light localization by graphene structure induces the change of the electron structure of molecules due to their direct interaction with the surface in the first adsorbed layers. However, there is no "chemical enhancement" and the one in SERS is associated with very strong change of the electric field, when one moves away from the surface [44]. In this work, G-NH$_2$/Au nanocomposite was used to identify the dye Rh6G and the neurotransmitter DA. DA is a catecholamine that plays a significant role in the functioning of central nervous, vascular, hormonal systems and its abnormal variation concentration in vivo has been linked to serious neurological diseases. The direct SERS quantification of DA in biological fluids remains a great challenge due to the low concentration (<10^{-10} M) and the high complexity of biological matrix. Raman features of Rh6G and DA were still detectable for concentration values down to 1×10^{-5} M and 1×10^{-6} M, respectively, although the sensibility of our system was found lower than the graphene-based SERS substrates reported in the literature for both analytes [24,45]. From our studies, it emerged that an improvement of G-NH$_2$/Au sensibility is imperative before proposing it as substrate for the detection of DA in biological matrix. We hypothesized that the detection limit of G-NH$_2$/Au nanocomposite could be improved by tuning the DA absorption properties on G and by setting the features, size, and shape of plasmonic noble metal NPs.

5. Conclusions

In summary, we investigated the SERS properties of a new graphene/gold nanocomposite (G-NH$_2$/Au) obtained by combining Au NPs produced by PLAL technique with G covalently functionalized (G-NH$_2$). After the chemical modification of G, the SERS platform was obtained by loading Au NPs on the G-NH$_2$ surface and deposition of G-NH$_2$/Au nanocomposite onto the glass slide by an aerography spraying technique. The chemical composition and the morphology of nanocomposites were investigated by micro-Raman XPS, STEM, and TGA analyses. STEM analyses showed transparent graphene sheets, with various dimensions, stacked onto each other, with a thickness of about 2–3 nm. Au NPs were detected as uniform spherical structures, with an average size of 15 nm, mainly distributed at the edges of the G layers. A good Au NPs loading was estimated by TGA and XPS analysis (i.e., 7.29% and 7.5%, respectively). This strategy allowed us to study SERS properties of G loaded with pure Au NPs without the influence of capping agents, surfactants, or salt produced in the chemical reduction of gold ions. SERS platform was tested to identify the dye Rh6G and the neurotransmitter DA; Raman features of Rh6G and DA are still detectable for concentration values down to 1×10^{-5} M and 1×10^{-6} M, respectively.

In conclusion, our platform possessed good stability and capability to reproduce the Raman signals without degradation although with low sensibility. Considering the feasibility of our method, further study will be devoted to improving the DA detection limits to refine the absorption properties of G-NH$_2$ and the plasmonic effect of loaded noble metal NPs.

Author Contributions: Conceptualization, A.P. and E.F.; methodology, G.N., E.F., P.G.M., A.S. and A.P.; validation, G.N., E.F., P.G.M., A.S. and A.P.; formal analysis, G.N., E.F., P.G.M., A.S. and A.P.; investigation, G.N., E.F. and P.G.M.; data curation, G.N., E.F., P.G.M., A.S. and A.P.; writing—original draft preparation, A.P. and E.F.; writing—review and editing, G.N., E.F, P.G.M., A.S. and A.P.

Funding: This research received no external funding.

Conflicts of Interest: The authors declare no conflict of interest.

References

1. Cataldi, P.; Athanassiou, A.; Bayer, I.S. Graphene Nanoplatelets-Based Advanced Materials and Recent Progress in Sustainable Applications. *Appl. Sci.* **2018**, *8*, 1438. [CrossRef]
2. Qu, Y.; He, F.; Yu, C.; Liang, X.; Liang, D.; Ma, L.; Zhang, Q.; Lv, J.; Wu, J. Advances on graphene-based nanomaterials for biomedical applications. *Mater. Sci. Eng. C* **2018**, *90*, 764–780. [CrossRef] [PubMed]
3. Reina, G.; González-Domínguez, J.M.; Criado, A.; Vázquez, E.; Bianco, A.; Prato, M. Promises, facts and challenges for graphene in biomedical applications. *Chem. Soc. Rev.* **2017**, *46*, 4400–4416. [CrossRef] [PubMed]
4. Silva, M.; Alves, N.M.; Paiva, M.C. Graphene-polymer nanocomposites for biomedical applications. *Polym. Adv. Technol.* **2018**, *29*, 687–700. [CrossRef]
5. Barreca, D.; Neri, G.; Scala, A.; Fazio, E.; Gentile, D.; Rescifina, A.; Piperno, A. Covalently immobilized catalase on functionalized graphene: Effect on the activity, immobilization efficiency, and tetramer stability. *Biomater. Sci.* **2018**, *6*, 3231–3240. [CrossRef] [PubMed]
6. Neri, G.; Micale, N.; Scala, A.; Fazio, E.; Mazzaglia, A.; Mineo, P.G.; Montesi, M.; Panseri, S.; Tampieri, A.; Grassi, G.; et al. Silibinin-conjugated graphene nanoplatform: Synthesis, characterization and biological evaluation. *FlatChem* **2017**, *1*, 34–41. [CrossRef]
7. Piperno, A.; Scala, A.; Mazzaglia, A.; Neri, G.; Pennisi, R.; Sciortino, M.T.; Grassi, G. Cellular Signaling Pathways Activated by Functional Graphene Nanomaterials. *Int. J. Mol. Sci.* **2018**, *19*, 3365. [CrossRef] [PubMed]
8. Neri, G.; Scala, A.; Barreca, F.; Fazio, E.; Mineo, P.G.; Mazzaglia, A.; Grassi, G.; Piperno, A. Engineering of carbon based nanomaterials by ring-opening reactions of a reactive azlactone graphene platform. *Chem. Commun.* **2015**, *51*, 4846–4849. [CrossRef] [PubMed]
9. Chen, Y.-W.; Su, Y.-L.; Hu, S.-H.; Chen, S.-Y. Functionalized graphene nanocomposites for enhancing photothermal therapy in tumor treatment. *Adv. Drug Deliv. Rev.* **2016**, *105*, 190–204. [CrossRef] [PubMed]

10. Wei, Y.; Zhou, F.; Zhang, D.; Chen, Q.; Xing, D. A graphene oxide based smart drug delivery system for tumor mitochondria-targeting photodynamic therapy. *Nanoscale* **2016**, *8*, 3530–3538. [CrossRef] [PubMed]
11. Wu, X.; Ding, S.-J.; Lin, K.; Su, J. A review on the biocompatibility and potential applications of graphene in inducing cell differentiation and tissue regeneration. *J. Mater. Chem. B* **2017**, *5*, 3084–3102. [CrossRef]
12. Suvarnaphaet, P.; Pechprasarn, S. Graphene-Based Materials for Biosensors: A Review. *Sensors* **2017**, *17*, 2161. [CrossRef] [PubMed]
13. Qiu, H.-J.; Guan, Y.; Luo, P.; Wang, Y. Recent advance in fabricating monolithic 3D porous graphene and their applications in biosensing and biofuel cells. *Biosens. Bioelectron.* **2017**, *89*, 85–95. [CrossRef] [PubMed]
14. Tao, G.; Wang, J. Gold nanorod@nanoparticle seed-SERSnanotags/graphene oxide plasmonic superstructured nanocomposites as an "on-off" SERS aptasensor. *Carbon* **2018**, *133*, 209–217. [CrossRef]
15. Hernández-Sánchez, D.; Villabona-Leal, G.; Saucedo-Orozco, I.; Bracamonte, V.; Pérez, E.; Bittencourt, C.; Quintana, M. Stable graphene oxide—Gold nanoparticle platforms for biosensing applications. *Phys. Chem. Chem. Phys.* **2018**, *20*, 1685–1692. [CrossRef]
16. Chisanga, M.; Muhamadali, H.; Ellis, D.I.; Goodacre, R. Enhancing Disease Diagnosis: Biomedical Applications of Surface-Enhanced Raman Scattering. *Appl. Sci.* **2019**, *9*, 1163. [CrossRef]
17. Yu, X.; Cai, H.; Zhang, W.; Li, X.; Pan, N.; Luo, Y.; Wang, X.; Hou, J.G. Tuning Chemical Enhancement of SERS by Controlling the Chemical Reduction of Graphene Oxide Nanosheets. *ACS Nano* **2011**, *5*, 952–958. [CrossRef]
18. Sitjar, J.; Liao, J.-D.; Lee, H.; Liu, B.H.; Fu, W.-E. SERS-Active Substrate with Collective Amplification Design for Trace Analysis of Pesticides. *Nanomaterials* **2019**, *9*, 664. [CrossRef]
19. Agarwal, N.R.; Fazio, E.; Neri, F.; Trusso, S.; Castiglioni, C.; Lucotti, A.; Santo, N.; Ossi, P.M. Ag and Au nanoparticles for SERS substrates produced by pulsed laser ablation. *Cryst. Res. Technol.* **2011**, *46*, 836–840. [CrossRef]
20. Lentini, G.; Fazio, E.; Calabrese, F.; De Plano, L.M.; Puliafico, M.; Franco, D.; Nicolò, M.S.; Carnazza, S.; Trusso, S.; Allegra, A.; et al. Phage–AgNPs complex as SERS probe for U937 cell identification. *Biosens. Bioelectron.* **2015**, *74*, 398–405. [CrossRef]
21. Harmsen, S.; Huang, R.; Wall, M.A.; Karabeber, H.; Samii, J.M.; Spaliviero, M.; White, J.R.; Monette, S.; O'Connor, R.; Pitter, K.L.; et al. Surface-enhanced resonance Raman scattering nanostars for high-precision cancer imaging. *Sci. Transl. Med.* **2015**, *7*, 271ra277. [CrossRef] [PubMed]
22. Marrucci, L.; Manzo, C.; Paparo, D. Optical Spin-to-Orbital Angular Momentum Conversion in Inhomogeneous Anisotropic Media. *Phys. Rev. Lett.* **2006**, *96*, 163905. [CrossRef] [PubMed]
23. Ningbo, Y.; Zhang, C.; Song, Q.; Xiao, S. A hybrid system with highly enhanced graphene SERS for rapid and tag-free tumor cells detection. *Sci. Rep.* **2016**, *6*, 25134. [CrossRef]
24. Silver, A.; Kitadai, H.; Liu, H.; Granzier-Nakajima, T.; Terrones, M.; Ling, X.; Huang, S. Chemical and Bio Sensing Using Graphene-Enhanced Raman Spectroscopy. *Nanomaterials* **2019**, *9*, 516. [CrossRef] [PubMed]
25. Moskovits, M.; Suh, J.S. Surface selection rules for surface-enhanced Raman spectroscopy: Calculations and application to the surface-enhanced Raman spectrum of phthalazine on silver. *J. Phys. Chem.* **1984**, *88*, 5526–5530. [CrossRef]
26. Xie, L.; Ling, X.; Fang, Y.; Zhang, J.; Liu, Z. Graphene as a Substrate to Suppress Fluorescence in Resonance Raman Spectroscopy. *J. Am. Chem. Soc.* **2009**, *131*, 9890–9891. [CrossRef]
27. Xu, W.; Xiao, J.; Chen, Y.; Chen, Y.; Ling, X.; Zhang, J. Graphene-Veiled Gold Substrate for Surface-Enhanced Raman Spectroscopy. *Adv. Mater.* **2013**, *25*, 928–933. [CrossRef]
28. Li, Y.; Yan, H.; Farmer, D.B.; Meng, X.; Zhu, W.; Osgood, R.M.; Heinz, T.F.; Avouris, P. Graphene Plasmon Enhanced Vibrational Sensing of Surface-Adsorbed Layers. *Nano Lett.* **2014**, *14*, 1573–1577. [CrossRef]
29. Ahmed, H.; Rezk, A.R.; Carey, B.J.; Wang, Y.; Mohiuddin, M.; Berean, K.J.; Russo, S.P.; Kalantar-Zadeh, K.; Yeo, L.Y. Ultrafast Acoustofluidic Exfoliation of Stratified Crystals. *Adv. Mater.* **2018**, *30*, 1704756. [CrossRef]
30. Mohiuddin, M.; Wang, Y.; Zavabeti, A.; Syed, N.; Datta, R.S.; Ahmed, H.; Daeneke, T.; Russo, S.P.; Rezk, A.R.; Yeo, L.Y.; et al. Liquid Phase Acoustic Wave Exfoliation of Layered MoS2: Critical Impact of Electric Field in Efficiency. *Chem. Mater.* **2018**, *30*, 5593–5601. [CrossRef]
31. Neri, G.; Scala, A.; Fazio, E.; Mineo, P.G.; Rescifina, A.; Piperno, A.; Grassi, G. Repurposing of oxazolone chemistry: Gaining access to functionalized graphene nanosheets in a top-down approach from graphite. *Chem. Sci.* **2015**, *6*, 6961–6970. [CrossRef] [PubMed]

32. Fazio, E.; Spadaro, S.; Santoro, M.; Trusso, S.; Lucotti, A.; Tommasini, M.; Neri, F.; Ossi, P. Synthesis by picosecond laser ablation of ligand-free Ag and Au nanoparticles for SERS applications. *EPJ Web Conf.* **2018**, *167*, 05002. [CrossRef]
33. Fazio, E.; Neri, F. Nonlinear optical effects from Au nanoparticles prepared by laser plasmas in water. *Appl. Surf. Sci.* **2013**, *272*, 88–93. [CrossRef]
34. Fazio, E.; Scala, A.; Grimato, S.; Ridolfo, A.; Grassi, G.; Neri, F. Laser light triggered smart release of silibinin from a PEGylated–PLGA gold nanocomposite. *J. Mater. Chem. B* **2015**, *3*, 9023–9032. [CrossRef]
35. Lavanya, N.; Fazio, E.; Neri, F.; Bonavita, A.; Leonardi, S.G.; Neri, G.; Sekar, C. Simultaneous electrochemical determination of epinephrine and uric acid in the presence of ascorbic acid using SnO_2/graphene nanocomposite modified glassy carbon electrode. *Sens. Actuators B Chem.* **2015**, *221*, 1412–1422. [CrossRef]
36. Wu, L.; Chu, H.S.; Koh, W.S.; Li, E.P. Highly sensitive graphene biosensors based on surface plasmon resonance. *Opt. Express* **2010**, *18*, 14395–14400. [CrossRef] [PubMed]
37. Zhang, H.-P.; Lin, X.-Y.; Lu, X.; Wang, Z.; Fang, L.; Tang, Y. Understanding the interfacial interactions between dopamine and different graphenes for biomedical materials. *Mater. Chem. Front.* **2017**, *1*, 1156–1164. [CrossRef]
38. Park, S.-K.; Lee, N.-S.; Lee, S.-H. Vibrational Analysis of Dopamine Neutral Base based on Density Functional Force Field. *Bull. Korean Chem. Soc.* **2000**, *21*, 959–968.
39. Murphy, S.; Huang, L.; Kamat, P.V. Reduced Graphene Oxide–Silver Nanoparticle Composite as An Active SERS Material. *J. Phys. Chem. C* **2013**, *117*, 4740–4747. [CrossRef]
40. Luo, Y.; Ma, L.; Zhang, X.; Liang, A.; Jiang, Z. SERS Detection of Dopamine Using Label-Free Acridine Red as Molecular Probe in Reduced Graphene Oxide/Silver Nanotriangle Sol Substrate. *Nanoscale Res. Lett.* **2015**, *10*, 230. [CrossRef]
41. Bottari, G.; Herranz, M.Á.; Wibmer, L.; Volland, M.; Rodríguez-Pérez, L.; Guldi, D.M.; Hirsch, A.; Martín, N.; D'Souza, F.; Torres, T. Chemical functionalization and characterization of graphene-based materials. *Chem. Soc. Rev.* **2017**, *46*, 4464–4500. [CrossRef] [PubMed]
42. Mishra, Y.K.; Adelung, R.; Kumar, G.; Elbahri, M.; Mohapatra, S.; Singhal, R.; Tripathi, A.; Avasth, D.K. Formation of Self-organized Silver Nanocup-Type Structures and Their Plasmonic Absorption. *Plasmonics* **2013**, *8*, 811–815. [CrossRef]
43. Norlander, P. The Ring: A Leitmotif in Plasmonics. *ACS Nano* **2009**, *3*, 488–492. [CrossRef] [PubMed]
44. Pockrand, I. *Surface Enhanced Raman Vibrational Studies at Solid/gas Interface*; Springer Tracts in Modern Physics; Springer: Berlin/Heidelberg, Germany; New York, NY, USA; Tokyo, Japan, 1984; Volume 104, pp. 1–164.
45. Yusoff, N.; Pandikumar, A.; Ramaraj, R.; Lim, H.N.; Huang, N.M. Gold nanoparticle based optical and electrochemical sensing of dopamine. *Microchim. Acta* **2015**, *182*, 2091–2114. [CrossRef]

© 2019 by the authors. Licensee MDPI, Basel, Switzerland. This article is an open access article distributed under the terms and conditions of the Creative Commons Attribution (CC BY) license (http://creativecommons.org/licenses/by/4.0/).

Article

Functionalization of Single and Multi-Walled Carbon Nanotubes with Polypropylene Glycol Decorated Pyrrole for the Development of Doxorubicin Nano-Conveyors for Cancer Drug Delivery

Chiara Pennetta [1], Giuseppe Floresta [2], Adriana Carol Eleonora Graziano [3], Venera Cardile [3], Lucia Rubino [1], Maurizio Galimberti [1], Antonio Rescifina [2,*] and Vincenzina Barbera [1,*]

[1] Department of Chemistry, Materials and Chemical Engineering "G. Natta", Politecnico di Milano, Via Mancinelli 7, 20131 Milano, Italy; chiara.pennetta@polimi.it (C.P.); luciarita.rubino@polimi.it (L.R.); maurizio.galimberti@polimi.it (M.G.)
[2] Department of Drug Sciences, University of Catania, Viale Andrea Doria 6, 95125 Catania, Italy; giuseppe.floresta@unict.it
[3] Department of Biomedical and Biotechnological Science, Section of Physiology, University of Catania, Via Santa Sofia 97, 95123 Catania, Italy; acegraz@unict.it (A.C.E.G.); cardile@unict.it (V.C.)
* Correspondence: arescifina@unict.it (A.R.); vincenzina.barbera@polimi.it (V.B.); Tel.: +39-095-738-5014 (A.R.); +39-02-738-4728 (V.B.)

Received: 9 March 2020; Accepted: 4 May 2020; Published: 31 May 2020

Abstract: A recently reported functionalization of single and multi-walled carbon nanotubes, based on a cycloaddition reaction between carbon nanotubes and a pyrrole derived compound, was exploited for the formation of a doxorubicin (DOX) stacked drug delivery system. The obtained supramolecular nano-conveyors were characterized by wide-angle X-ray diffraction (WAXD), thermogravimetric analysis (TGA), high-resolution transmission electron microscopy (HR-TEM), and Fourier transform infrared (FT-IR) spectroscopy. The supramolecular interactions were studied by molecular dynamics simulations and by monitoring the emission and the absorption spectra of DOX. Biological studies revealed that two of the synthesized nano-vectors are effectively able to get the drug into the studied cell lines and also to enhance the cell mortality of DOX at a much lower effective dose. This work reports the facile functionalization of carbon nanotubes exploiting the "pyrrole methodology" for the development of novel technological carbon-based drug delivery systems.

Keywords: carbon nanotubes; pyrrole; cancer; doxorubicin; drug delivery systems

1. Introduction

sp^2 carbon allotropes are fascinating materials. The discovery of fullerenes [1] "stimulated the creativity and imagination of scientists and paved the way to whole new chemistry and physics of nanocarbons" [2]. Within these materials, carbon black [3] is one of the ten most important chemical products, and recently novel families of sp^2 carbon allotropes have become of great interest: carbon nanotubes), both single (SWCNT) [3,4] and multi-walled (MWCNT) [5,6], graphene [7–10] or graphitic nanofillers made by few layers of graphene [11–14].

CNT have a peculiar combination of electrical, thermal, and mechanical properties [15–18]. CNT find applications as superconductors [19], electrochemical capacitors [20], electromechanical actuators [21], photovoltaic devices [22,23], nanowires [24], in nanocomposite materials [25–27] and in medicinal chemistry [28]. CNT are emerging nanomaterials with great potential for diagnostic and therapeutic applications in medicine [28–30]. In this field, the biocompatibility of CNT is a hugely important aspect that has to be considered. In this direction, Prato and colleagues have reported some basic rules in the design of functionalized CNT to avoid deposition in specific tissues [28]. In particular, it was revealed that functionalized SWCNT are degraded inside cells such as neutrophils and macrophages in relation to the type of functional groups chemically introduced on the surface of CNT [31].

In the design of a potential CNT-based carrier, it is important to consider that: (i) they are usually available as highly entangled bundles; (ii) their dispersion in a matrix or a solvent is indeed very difficult. Moreover, the compatibility of CNT with the matrix depends on the solubility parameters of tubes and matrix. An effective way to overcome these problems is the functionalization of CNT, and several ways of achieving this have thus been developed. Functionalization of sp^2 carbon allotropes can be classified as covalent [32,33] and non-covalent [34,35]. Some authors reported a sustainable functionalization method that is based on the principles of green chemistry [36], based on easily available chemicals, ideally biosourced [36,37] and economically feasible.

Moreover, the main goal was to identify a method able to promote the functionalization of most, if not all, the families of sp^2 carbon allotropes. Graphene layers have been functionalized with pyrrole compounds obtained from the *Paal Knorr* reaction of a primary amine with a diketone, 2,5-hexanedione (HD). The selected sp^2 carbon allotrope was functionalized by simply mixing it with the pyrrole compound (PyC) and giving either mechanical or thermal energy [37]. In the latter case, it was shown [37] that the graphene layers constituting the bulk structure of the carbon substrate remained substantially unaltered. Very high functionalization yields were reported [37]; in some cases, the functionalization achieved was from 80% to 95%. Studies on the reaction mechanism led to the hypothesis [38] that covalent bonds are formed between the carbon substrate and the pyrrole compound, with the occurring of a domino reaction: carbocatalyzed oxidation of the pyrrole compound and subsequent *Diels-Alder* cycloaddition. Preliminary indications have been reported [39] that the functionalization with different pyrrole compounds leads to the modification of the solubility parameter of graphene layers in a broad range of values.

This manuscript reports on the functionalization of single and multiwalled carbon nanotubes from now on indicated as CNT with a pyrrole compound obtained from HD and an amine-terminated poly(propylene glycol). Synthesis of the *O*-(2-(2,5-dimethyl-1*H*-pyrrol-1-yl) propyl)-*O'*-(2-methoxyethyl)polypropylene glycol (pyrrole polypropylene glycol, PPGP) is presented, the functionalization of CNT is described, and the main characteristics of the CNT/PPGP adducts are discussed. Two different adducts were prepared: the supramolecular adduct (CNT/PPGP$_s$) and the covalent adduct (CNT/PPGP$_c$). Characterization of the adducts was carried out employing thermogravimetric analysis (TGA), wide-angle X-ray diffraction (WAXD), high-resolution transmission electron microscopy (HR-TEM), and Fourier transform infrared spectroscopy (FT-IR). The ability of PyC to modify the solubility parameter of CNT was investigated. Based on the solubility parameter study and the easy functionalization procedure, the pyrrole ring was selected as the reactive moiety. The presence of the pyrrole at the end of the polymer chain allows obtaining a better conjugation with CNT respect to the polyether chain without a functionalizing molecule at the end.

The formation of the ternary nano complexes carbon nanotubes/polypropylenglycolpyrrole/ doxorubicin (CNT/PPGP/DOX) is also reported. The supramolecular interactions between CNT/PPGP and DOX were studied by monitoring the emission and the absorption spectra of the drug employing fluorescence and Ultraviolet-visible (UV-Vis) spectrophotometries, respectively, and by molecular dynamics simulations. The ability of DOX to interact non-covalently with pristine and functionalized CNT and the evaluation of their capability to kill human melanoma and lung cancer cells is also reported. The release of DOX at slightly acidic pH was observed to be fast for MWCNT/PPGPc and slow for MWCNT/PPGPs and MWCNT. Preliminary biological studies revealed that two of the synthesized nano-vectors are effectively able to release the drug in situ.

2. Experimental Part

2.1. Materials

Reagents and solvents are commercially available and were used without any further purification: Methanol, 2-propanol, ethyl acetate, propylene glycol, dichloromethane, xylene, toluene, n-hexane, DOX hydrochloride, O-(2-aminopropyl)-O'-(2-methoxyethyl)polypropylene glycol (PPGA) (average $M_n \approx 600$) and deuterated chloroform (CDCl$_3$) were purchased from Sigma-Aldrich Merck KGaA group, (Darmstadt, Germany). Carbon nanotubes (CNT) were the sp^2 carbon allotropes used in this work: multi-wall carbon nanotubes (MWCNT) were NANOCYL® NC7000™ series from Nanocyl SA (Sambreville, Belgium), with a carbon purity of 90%, average length of about 1.5 µm and BET surface area (Brunauer Emmett Teller surface area) of 275 m^2/g; single-wall carbon nanotubes (SWCNT) were TUBALL™ from OCSiAl (Leudelange, Grand-Duché de Luxembourg) with a carbon purity > 85%, average length of about 5 µm and BET surface area of 332 m^2/g.

2.2. Synthesis of O-(2-(2,5-Dimethyl-1H-Pyrrol-1-Yl) Propyl)-O'-(2-Methoxyethyl)Polypropylene Glycol (Pyrrole Polypropylene Glycol, PPGP)

A quantits of 13.27 g of O-(2-Aminopropyl)-O'-(2-methoxyethyl)polypropylene glycol (PPGA, 0.02212 mol) and 2 g of 2,5-hexanedione (HD, 0.02212 mol) were poured in a 100 mL round-bottomed flask equipped with a magnetic stirrer. The mixture was then stirred (300 rpm) at 100 °C for 4 h. At the end of the reaction, the mixture still contained 2,5-hexanedione. The reagent was removed at reduced pressure (2 mbar, 25 °C) by using a Claisen apparatus (Colaver s.r.l, Vimodrone, Italy). The pure product was obtained with a yield of 64%. ^{1}H NMR (CDCl$_3$, 400 MHz); δ (ppm) = 1.13–1.15 (m, 74H); 1.47 (d, 3H); 2.25 (s, 6H); 3.32 (m, 4H); 3.32 (m, 4H); 3.42–3.78 (m, 34H); 3.66–3.5 (m, 55H); 3.74 (m, 3H); 4.35 (quintet, 1H); 5.68 (s, 2H); ^{13}C NMR (CDCl$_3$, 100 MHz); δ (ppm) = 127.0, 105.0, 80, 71.2, 70.0, 68.3, 62.0, 59.3, 31.0, 20.0, 13.2.

2.3. Preparation of CNT/PPGP Adducts

2.3.1. Preparation of CNT/PPGP Supramolecular Adduct (CNT/PPGP$_s$)

In a 250 mL round-bottomed flask CNT (500 mg) and acetone (20 mL) were put in sequence. The system was sonicated for 30 min, and then the pyrrole derivative (150 mg) was added into the flask. The suspension was sonicated again for 30 min. After solvent removal under reduced pressure, the mixture was quantitatively transferred in a funnel with a sintered glass disc, washed with acetone (100 mL), and then recovered and weighed.

The degree of functionalization was estimated employing TGA, determining the amount of pyrrole compound in the adduct after washing (mass losses for CNT and CNT/PPGP adducts are in Table 1), see Table 2: SWCNT/PPGP$_s$ 5.48 w%; MWCNT/PPGP$_s$ 3.00 w%.

Table 1. Mass losses for CNT and CNT/PPGP adducts from TGA analysis.

Samples	Mass Loss (%)		
	0 < T < 150 °C	150 < T < 700 °C	700 < T < 900 °C
SWCNT [a]	1.6	1.0	97.4
SWCNT/PPGP$_s$ [a,b]	2.2	5.4	92.4
SWCNT/PPGP$_c$ [a,c]	0.1	5.0	94.9
MWCNT [d]	2.0	0.2	97.8
MWCNT/PPGP$_s$ [b,d]	0.2	3.5	96.3
MWCNT/PPGP$_c$ [c,d]	0.1	8.0	91.9

[a] Single-wall carbon nanotube; [b] supramolecular adduct; [c] covalent adduct; [d] multiwall carbon nanotubes.

Table 2. Degree [a] and yield [b] of functionalization of CNT functionalized with PPGP.

Adduct	Degree of Functionalization (%)	Functionalization Yield (%)
SWCNT/PPGP$_s$	5.5	75
SWCNT/PPGP$_c$	5.0	70
MWCNT/PPGP$_s$	3.0	56
MWCNT/PPGP$_c$	7.0	84

[a] Calculated through Equation (1); [b] calculated through Equation (2).

2.3.2. Preparation of CNT/PPGP Covalent Adduct (CNT/PPGP$_c$)

In a 250 mL round-bottomed flask equipped with magnetic stirrer, we added CNT (500 mg) and 20 mL of acetone in sequence. The system was sonicated for 30 min, and then the pyrrole derivative (150 mg) was added into the flask. The suspension was sonicated again for 30 min. After solvent removal under reduced pressure, the CNT/PPGP mixture was poured in a round bottom flask and heated at 150 °C for 2 h. After this time, the mixture was quantitatively transferred in a funnel with a sintered glass disc, washed with acetone (100 mL), and then recovered and weighed.

The degree of functionalization was estimated employing TGA, determining the amount of pyrrole compound in the adduct after washing (mass losses for CNT and CNT/PPGP adducts are in Table 1), see Table 2: SWCNT/PPGP$_c$ 5 w%; MWCNT/PPGP$_c$ 7.00 w%.

2.4. Preparation of Carbon Nanotube/Pyrrole Polypropylene Glycol/Doxorubicin CNT/PPGP/DOX Ternary Nano Complexes

2.4.1. General Procedure

DOX hydrochloride (9 mg) was stirred with the selected CNT/PPGP adduct (3 mg) dispersed in a pH 7.4 Phosphate Buffered Saline solution (PBS) (6 mL) and stirred for 16 h at room temperature. The product was collected by ultracentrifugation with PBS until the supernatant became colorless. The amount of unbound DOX was determined by measuring the absorbance at 490 nm of the supernatant after centrifugation see Table S1. The CNT/PPGP/DOX nano complex dispersions in PBS were also analyzed by fluorescence spectrophotometry: dispersions were placed using a Pasteur pipette (Colaver s.r.l, Vimodrone, Italy) in a triangular quartz cuvette. A Jasco FP-6600 Spectrofluorometer (JASCO corporation, Tokyo, Japan) was used to perform the fluorescence measurements. The fluorescence detector was set at an excitation wavelength of 480 nm, and fluorescence spectra in the range of 500–700 nm were collected.

2.4.2. DOX Calibration Curve by UV-Vis Spectroscopy

A stock PBS solution of DOX hydrochloride (1 mg/mL) at pH 7.4 was prepared. The obtained solution was then diluted, and UV-Vis measurements were performed. The absorbance of these solutions was measured at 490 nm of maximum absorbance, using a 1 cm quartz cuvette.

2.4.3. DOX Release From CNT Nano Complexes

CNT/DOX and CNT/PPGP/DOX in acetate buffer (pH 5.5) were loaded in Spectra/Por® Dialysis Membrane (10K MWCO, nominal flat width 24 mm, diameter 15 mm wet in 0.1% sodium azide) (Thermo Fisher Scientific Inc., Waltham, MA, USA). Each dialysis bag was then allowed to stand for 72 h in acetate buffer solution. The release of DOX was checked after 24, 48, and 72 h through UV-Vis spectroscopy. The same experiment conducted in PBS showed that DOX remained bound to CNT due to the stability of the drug at pH 7.4.

2.5. Characterization of Pristine CNT, CNT/PPGP Adducts and CNT/PPGP/DOX Ternary Nano Complexes

2.5.1. Fourier Transform Infrared Spectroscopy (FT-IR)

The IR spectra were recorded in transmission mode (128 scans and 4 cm^{-1} resolution) using a Thermo Electron Continuum IR microscope coupled with an FTIR Nicolet Nexus spectrometer. A small portion of the dry solid material was placed in a diamond anvil cell (DAC) and analyzed in transmission mode.

2.5.2. Thermogravimetric Analysis (TGA)

TGA test under N_2 flowing (60 mL/min) was performed with a Mettler TGA SDTA/851 instrument according to the ISO9924-1 standard method. Samples (10 mg) were heated from 30 to 300 °C at 10 °C/min, kept at 300 °C for 10 min, and then heated up to 550 °C at 20 °C/min. After being maintained at 550 °C for 15 min, they were further heated up to 700 °C and kept at 700 °C for 30 min under air flowing (60 mL/min).

2.5.3. High-Resolution Transmission Electron Microscopy (HR-TEM)

HR-TEM investigations on CNT and CNT/PPGP adducts were carried out with a Philips CM 200 field emission gun microscope operating at 200 kV. Few drops of the water suspensions were deposited on 200 mesh lacey carbon-coated copper grid and air-dried for several hours before analysis. During the acquisition of HR-TEM images, performed with low beam current densities and short acquisition times, the samples did not undergo structural transformation. The Gatan Digital Micrograph software (GMS 3, Gatan, Inc., Pleasanton, CA, USA) was used to estimate in HRTEM micrographs the number of stacked graphene layers and the dimensions of the stacks.

2.5.4. Wide-Angle X-Ray Diffraction

Wide-angle X-ray diffraction patterns were obtained in reflection, with an automatic Bruker D8 Advance diffractometer (Bruker Corporation, Billerica, MA, USA), with nickel filtered Cu–K$_\alpha$ radiation. Patterns were recorded in 10–80° as the 2θ range, being 2θ the peak diffraction angle. Details can be found in Section S1.

2.6. Preparation and Characterization of Dispersions of CNT/PPGP Adducts in Different Solvents-Evaluation of Solubility Parameters

2.6.1. Preparation of Water Dispersions of CNT/PPGP Adducts

General procedure: water dispersions of CNT/PPGP adducts were prepared at different concentrations: 1 mg/mL; 0.5 mg/mL; 0.1 mg/mL; 0.05 mg/mL; 0.01 mg/mL; 0.005 mg/mL; 0.001 mg/mL. Each dispersion was sonicated for 1 min using an ultrasonic bath (260 W). The dispersion (10 mL) of each sample was put in a Falcon™ 15 mL Conical Centrifuge Tubes and centrifuged at 6000 rpm for 30 min. UV-Vis measurement was performed immediately after sonication or centrifugation, and also after 3 days. A Hewlett Packard 8452A Diode Array Spectrophotometer (Hewlett-Packard, Palo Alto, CA, USA) was used to perform the absorption measurements. The dispersions were placed using a Pasteur pipette in cuvettes with an optical path of 1 cm (about 3 mL per cuvette). The obtained UV-visible spectrum reports absorption as a function of radiation wavelength in the range of 200–750 nm.

2.6.2. Calculation of the Hansen Solubility Sphere and Hansen Solubility Parameters

The calculation of the *Hansen* solubility parameters (HSP) for CNT was performed by applying the Hansen Solubility Sphere representation of miscibility. The idea at the basis of this geometrical approach is the calculation of the cohesive energy density (U_T/V) of a compound as the sum of three interaction contributions: non-polar Van der Waals forces (δ_D), polar (δ_P) and hydrogen bonding (δ_H). Details can be found in Section S2.

2.7. Biological Studies

2.7.1. Cell Cultures

The M14 human melanoma and the A549 human lung adenocarcinoma cell lines have been used in the present investigation. Both cell lines were routinely maintained as previously described [40,41] in a humidified atmosphere of 5% CO_2 in a water-jacketed incubator at 37 °C. The base media were Roswell Park Memorial Institute (RPMI) 1640 for M14 cells, and Dulbecco's modified Eagle's medium (DMEM) for A549 cells. The complete growth media were obtained by supplementation with 10% (by volume) heat-inactivated fetal bovine serum, 2 mM glutamine, 100 units/mL penicillin, and 100 µg/mL streptomycin (Thermo Fisher Scientific, Milan, Italy). The cells were subcultured before reaching confluence, using a 0.25% trypsin-EDTA solution (Carlo Erba, Milan, Italy).

2.7.2. Cell Viability Assay

Cells were seeded in 96-well cell culture plates at a density of 2×10^4 cells/well (200 µL/well) for in vitro cell viability assay. After overnight incubation to allow the attachment of cells, the resulting monolayers were incubated with free DOX hydrochloride or into both unloaded and loaded CNT for 48 h.

In these experiments, all CNT samples were bath sonicated for 5 min in culture medium in order to obtain a homogeneous dispersion at 1 mg/mL. The stock dispersions were diluted in culture medium to get the desired concentrations referred and normalized to the amount of loaded DOX hydrochloride into each sample. After the time of incubation, the cells were carefully washed to remove the non-internalized nanotubes, and cytotoxicity was determined by 3-(4,5-dimethyl-2-thiazolyl)-2,5-diphenyl-2H-tetrazolium bromide (MTT) assay according to a previously published protocol [42]. The formazan was solubilized in DMSO (Sigma-Aldrich, Italy) and spectrophotometrically quantified at $\lambda = 550$ nm by a microplate reader (Titertek Multiscan, DAS, Milan, Italy).

2.7.3. Statistical Analysis

The results were expressed as mean ± s.e.m. based on data derived from three independent experiments run in triplicate. Statistical analysis of results was performed using Student's t-test and one-way ANOVA test plus Dunnett's test by the statistical software package SYSTAT, version 11 (Systat Inc., Evanston, IL, USA).

2.8. Computational Details

2.8.1. Models Preparation

SWCNT and MWCNT were generated from the Nanotube Modeler package [43] (v. 1.8, JCrystalSoft) with the armchair arrangement, all open and hydrogen-terminated, trying to be faithful, as diameter and number of walls, to the values reported in the literature [34]. The SWCNT structure with a length of 50 Å and the chiral vectors m = 13, and n = 13 (13,13) with a diameter of 17.640 Å, and the MWCNT structure with a length of 10 Å and the chiral vectors (25,25) for an inner diameter of 33.924 Å and (70,70) for an outer diameter of 94.987 Å were used as the model for the CNT/PPGP/DOX drug carrier systems. The initial structure of DOX was obtained from the DRUGBANK server [44], whereas that of PPGP was manually built.

2.8.2. Molecular Dynamics Simulations

The molecular dynamics simulations of the CNT/PPGP/DOX supramolecular systems (SSy) were performed with the YASARA Structure package (v. 19.10.23). A periodic simulation cell with boundaries extending 10 Å [45] from the surface of the CNT or CNT/PPGP$_c$ was employed. For the two SWCNT systems (1066 carbon atoms), we used 1 PPGP and 45 DOX molecules (SSy1,2, see paragraph 3.4), whereas for the two MWCNT systems (8550 carbon atoms), we used 12 PPGP and 369 DOX molecules for PPGP$_s$ arrangement, and 5 PPGP and 354 DOX molecules for the PPGP$_c$ one (SSy3,4, see paragraph 3.4). For the SWCNT/PPGP$_c$ system, the one PPGA molecule was covalently linked about at the center of the CNT structure, whereas for the MWCNT/PPGP$_c$ system, the five PPGA molecules were covalently linked, approximately regularly spaced, around the outer circumference of the CNT structure. Details can be found in Section S3.

3. Results and Discussion

3.1. Modification of the CNT with a Pyrrole Decorated Polypropylene Glycol

Functionalization of carbon nanotubes, either single (SWCNT) and multi-walled (MWCNT), was realized by using a pyrrole terminated polymer (Pyrrole polypropylene glycol, PPGP) as a modifying agent. PPGP was synthesized through the *Paal-Knorr* reaction [34,39], by reacting the amino terminated polymer O-(2-Aminopropyl)-O'-(2-methoxyethyl)polypropylene glycol (PPGA) with HD as reported in the experimental part and summarized in Figure 1.

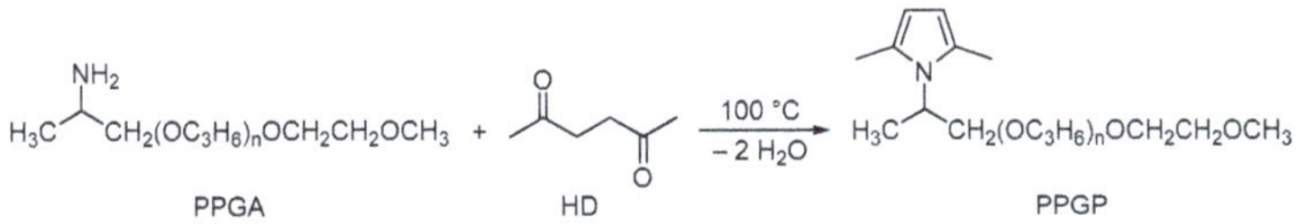

Figure 1. Synthesis of O-(2-(2,5-dimethyl-1H-pyrrol-1-yl) propyl)-O'-(2-methoxyethyl)polypropylene glycol (Pyrrole polypropylene glycol, PPGP).

The reaction was performed in the absence of catalysts by adopting experimental conditions inspired by the basic principles of green chemistry [36]. In particular, acidic catalysts and toxic solvents traditionally used for the reactions of primary amines with carbonyl compounds were avoided. In previous works [34,37], it was shown that the reaction of 2-amino-1,3-propanediol (serinol) with 2,5-hexanedione led to a 1,3-bis-oxazolidine compound, which was then converted into a pyrrole compound only by heating. The reaction of PPGA with HD was performed by heating at 100 °C. Water was the only co-product of the reaction, which was characterized by a high atom economy of 88.2% and by a yield of 64%. The chemical structure of PPGP was confirmed through ^{1}H NMR spectroscopy Figure S1. PPGP appears as a light-yellow liquid with slightly higher viscosity at room temperature and normal pressure. PPGP contains a long polyether chain that could favor the compatibility with polar surroundings such as water, alcohols, and polar polymer matrices. PPGP also contains the pyrrole ring, which could favor π–π interaction with the aromatic rings of the sp^2 carbon allotropes.

As stressed in the introduction, pyrrole compounds (PyC) were shown to be able to form stable adducts with graphitic substrates [37,38]. In this work, reactions between CNT and PPGP were performed by adopting the experimental frame summarized in Figure 2.

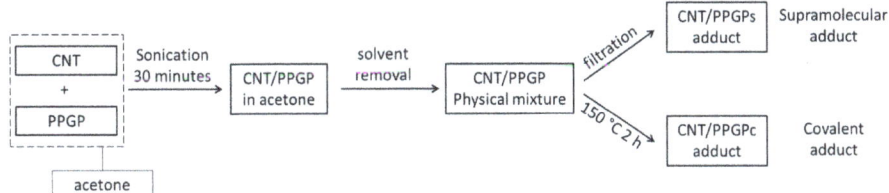

Figure 2. Block diagram summarizing the preparation of CNT/PPGP adducts.

In the experimental part, the functionalization of CNT by using PPGP as the modifying agent is described extensively. In brief, CNT were initially sonicated in the presence of the pyrrole terminated polymer (PPGP) in acetone, leading to the formation of the CNT/PPGP$_s$ adducts (supramolecular adduct) after removal of the solvent and washing with acetone. However, in a one-pot process involving the direct thermal treatment of the physical mixture (150 °C, 2 h), the covalent adduct CNT/PPGP$_c$ was easily obtained. No further optimization of the reaction conditions was performed.

In order to establish the efficiency of the functionalization, the CNT/PPGP powder taken from the flask was extracted in *Soxhlet* with acetone until PPGP was undetectable in the washing solvent. TGA was performed on CNT and the CNT/PPGP adducts before and after washing with acetone. As described in detail in the experimental part, the TGA was carried out under nitrogen up to 700 °C and under oxygen up to 800 °C. Thermographs of both single and multiwall CNT and CNT/PPGP adducts are shown in Figure 3. The values of mass losses are reported in Table 1.

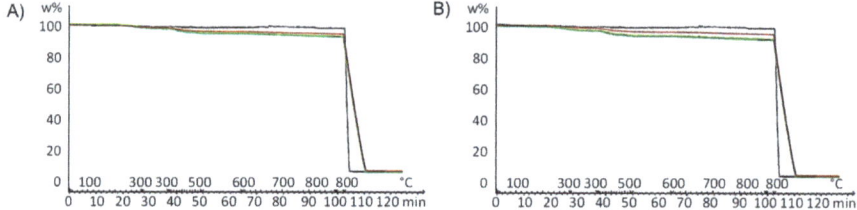

Figure 3. Thermogravimetric analysis (TGA) curves of: (**A**) single-wall carbon nanotubes (SWCNT) (black line), SWCNT/PPGP$_s$ (green line), and SWCNT$_c$ (red line); (**B**) multi-walled carbon nanotubes(MWCNT) (black line), MWCNT/PPGP$_s$ (green line), and MWCNT$_c$ (red line).

The mass loss at T < 150 °C was attributed to absorbed low molar mass molecules, mainly water. The decomposition profile for all the samples, pristine CNT and CNT/PPGP adducts, reveals two main steps in the temperature range from 150 °C to 700 °C, which can be attributed to the decomposition of alkenyl-, oxygen- and nitrogen-containing functional groups. New final decomposition occurs at T > 700 °C, due to reaction with the oxygen of the graphitic structure. The amount of PPGP in the adduct was estimated by evaluating the mass loss in the temperature range from 150 °C to 700 °C.

By comparing data referring to CNT and CNT/PPGP adducts, it appears that the mass loss is more significant for the latter ones. It can also be observed that CNT/PPGP adducts containing the polymeric chain give rise to a more substantial mass loss below 150 °C, and this could be due to a more considerable amount of absorbed water.

The degree of functionalization and the functionalization yield were calculated through Equations (1) and (2), and their values are reported in Table 2.

$$\text{Degree of functionalization } (\%) = 100 \cdot [(wt.\% \text{ loss } CNT/PPGP) - (wt.\% \text{ loss } CNT)] \quad (1)$$

$$\text{Functionalization yield } (\%) = 100 \cdot \frac{PPGP \text{ mass } \% \text{ in } (CNT/PPGP \text{ adduct}) \text{ after acetone washing}}{PPGP \text{ mass } \% \text{ in } (CNT/PPGP \text{ adduct}) \text{ before acetone washing}} \quad (2)$$

Although optimization of the reaction conditions was not performed, functionalization yield was high for all the CNT adducts except for the MWCNT/PPGP$_s$ adduct, which was 56%. This significant difference in the functionalization efficiency between the covalent and the supramolecular MWCNT adducts occurs since the commercial MWCNT are usually much entangled. The covalent functionalization helps the disentanglement of MWCNT increasing the available surface, which allows a better interaction between the functionalizing molecule and the MWCNT. In the MWCNT supramolecular adducts, there is no such effect, and the carbon nanotubes remain much entangled.

Mass loss values in the range 150–900 °C were exploited to determine the degree of functionalization of each adduct, defined as the percent value of the difference between the mass loss of the CNT/PPGP and that of the pristine CNT in the studied temperature range.

Thermogravimetric experiments showed that the functionalization of the considered adduct with PPGP was successful and that the introduction of an appropriate amount of modifier was achieved. The characterization of the CNT/PPGP adducts was performed employing IR and WAXD spectroscopies.

Infrared spectroscopy allowed a qualitative check of the chemical nature of the attached molecule and was performed on pristine CNT and CNT/PPGP adducts after acetone extraction. In Figure 4A,B the IR spectra of CNT and CNT/PPGP adducts are reported. As explained in detail in the experimental part, IR spectra have been recorded in transmission mode using a diamond anvil cell (DAC) in order to avoid the absorption peaks of water molecules. The spectra were obtained from the absorption of very thin films of CNT powder, which are not transparent to the IR beam. Indeed, the G_{IR} absorption observed in the spectra at 1590 cm^{-1} is mostly due to the reflection from the graphitic planes. The intense light diffusion from the high surface area graphite (HSAG) particles is responsible for the increase of the absorbance in the spectra toward higher wavenumbers. Absorption spectra recorded on DAC are presented in the region 700–3900 cm^{-1} Figure 4A and in the fingerprint region 700–1800 cm^{-1} Figure 4B after baseline correction to make easier the comparison of the weak spectroscopic features. The low and broad vibrational signals floating on a rather steep background absorption and the chemical complexity of the samples severely limits a structural diagnosis through a detailed assignment of the spectral features. Thus, the vibrational analysis was based on the recognition of the functional groups based on correlative spectroscopic criteria [38]. The FT-IR spectra of MWCNT, MWCNT/PPGP$_s$, and MWCNT/PPGP$_c$ are reported in Figure 4. In particular, Figure 4A shows the spectra recorded in the region 700–3900 cm^{-1}, while in Figure 4B the spectra are displayed in the fingerprint region, after baseline correction, to allow easier comparison. In Figure 4, the IR spectra of MWCNT (a) MWCNT/PPGP$_s$ (b) and MWCNT/PPGP$_c$ (c) are characterized by the presence of the typical strong feature at 1590 cm^{-1}. The spectra of MWCNT/PPGP$_s$ and MWCNT/PPGP$_c$ are both dominated by the very strong absorptions characteristic of their functional groups. Both spectra are dominated by (i) the absorption near 1100 cm^{-1} assigned to the stretching of C–O–C groups and by (ii) the broad bands at 1650 cm^{-1} and 1400 cm^{-1} assigned to the C=C stretching of the pyrrole ring. All these vibrations are compatible with the presence of PPGP functionalities. In the spectrum of MWCNT/PPGP$_s$, bands related to unreacted pyrrole compounds are present: C–N stretching (aromatic amines) from 1335–1250 cm^{-1}.

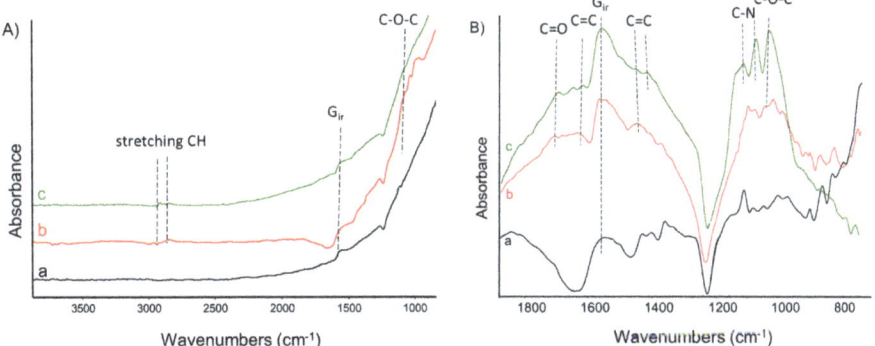

Figure 4. FT-IR spectra of MWCNT (a), MWCNT/PPGP$_s$ (b), and MWCNT/PPGP$_c$ (c). (**A**): 700–3900 cm^{-1} region; (**B**) fingerprint region 700–1800 cm^{-1} after baseline correction. Spectra are displayed with normalized intensity. Peaks discussed in the text are labeled.

As reported in the introduction, another aim of this study was to investigate the ability of PPGP to modify the solubility parameter of carbon nanotubes in order to preliminarily understand if the projected carrier could be able to interact with different surrounding. Dispersions of CNT/PPGP adducts cited in Table 1 were prepared in solvents having different solubility parameters: water, methanol, 2-propanol, acetone, ethyl acetate, propylene glycol, dichloromethane, xylene, toluene, and hexane. The stability of such dispersions was studied, as described in the experimental part. Visual inspection of the dispersions was carried out immediately after sonication. In Table S2, the result of these observations is qualitatively summarized with 'good' (meaning that a homogenous dispersion was observed) or 'bad' (the adduct either settled down or floated on the solvent) as indicators.

The introduction of PPGP on the surface of the carbon allotropes allows the dispersion in polar environments thanks to the long polyether chain. The solubility sphere shown in Figure 5 was generated, as explained in the experimental part, to encompass the suitable solvents points and to exclude the wrong solvents, being centered on the solubility parameters of the MWCNT/PPGP$_c$ adduct. Solubility parameters, δ_D, δ_P, and δ_H values, of MWCNT/PPGP$_c$ adduct, were estimated to be 11.48 MPa$^{0.5}$, 15.40 MPa$^{0.5}$, and 18 MPa$^{0.5}$, respectively.

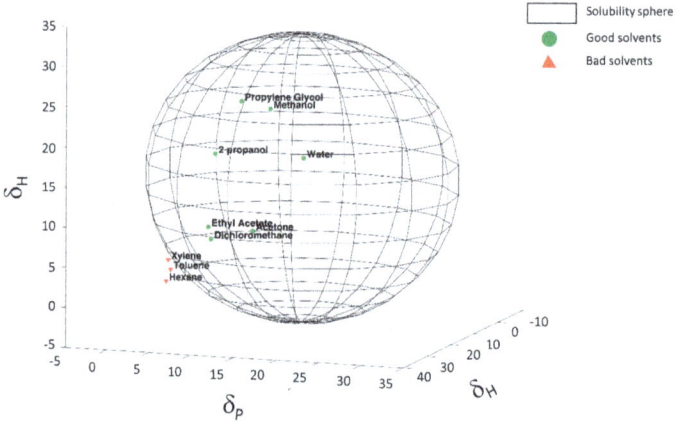

Figure 5. *Hansen* solubility sphere calculated for MWCNT/PPGP$_c$ adduct. The green circles correspond to the suitable solvents (within the radius of interaction), the red triangles to the wrong solvents (outside the sphere).

As it should be expected, considering the presence of such functional groups and a polyether long chain, water dispersions of CNT/PPGP adducts were easily prepared. Preparation and UV-Vis absorption data of water dispersions of CNT/PPGP adducts were reported and discussed in the following part.

CNT/PPGP water dispersions with the following concentrations were prepared: 1, 0.5, 0.1, 0.05, 0.01, 0.005 and 0.001 mg/mL. The dispersions were then analyzed by observing their absorptions in the UV-Vis range. Details are reported in the experimental section for each CNT/PPGP adduct. The adopted procedure is summarized in the block diagram of Figure 6.

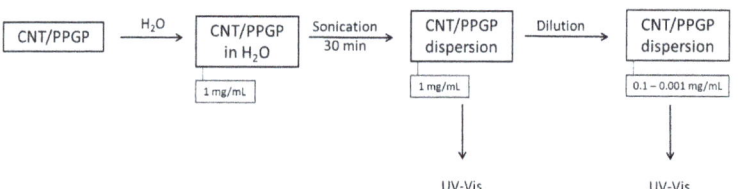

Figure 6. Block diagram summarizing the preparation procedure and characterization protocol adopted for UV-Vis analysis on CNT/PPGP adducts dispersions at several concentrations.

Adduct dispersions at different concentrations were prepared to investigate if the Lambert-Beer law is respected. If so, the CNT/PPGP adduct can be assumed to form a "solution-like" substance with water, thanks to the added polyether chain on its surface. Figure 7 shows the dependence of UV-Vis absorbance on the concentration of MWCNT/PPGP$_c$ adduct in water after sonication (A) and the linear relationship between absorbance at 260 nm and concentration for MWCNT/PPGP$_c$ (B).

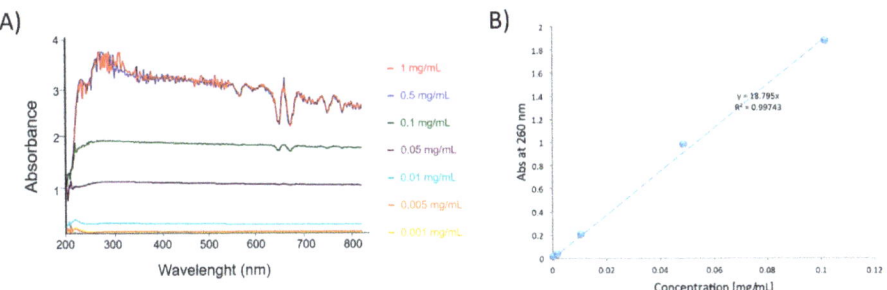

Figure 7. (**A**) Dependence of UV-Vis absorbance on the concentration of MWCNT/PPGP$_c$ adduct in water after sonication; (**B**) Linear relationship between absorbance at 260 nm and concentration for MWCNT/PPGP$_c$ adduct.

Furthermore, curves related to the 0.005 and 0.001 mg/mL dispersions reveal the presence of a non-negligible absorbance; this suggests high stability of the functionalized fractionated adduct still in suspension, not observed in the case of non-functionalized CNT.

Absorption values corresponding to a wavelength of 260 nm for MWCNT-PPGP$_c$ adduct curves were plotted as a function of concentration values. In Figure 7B, a linear correlation is reported.

Structure and morphology of the CNT and the ensuing CNT/PPGP adducts were investigated through WAXD and HR-TEM analysis. Figure S4 shows WAXD patterns for powders of MWCNT, MWCNT/PPGP$_s$, and MWCNT/PPGP$_c$. In pristine MWCNT Figure S4a, crystalline order in the direction orthogonal to structural layers is revealed by 002 reflection at 26.6°, which corresponds to an interlayer distance of 0.35 nm. By applying the *Scherrer* equation (Equation (S2)) to (002) reflection, the out of the plane ($D_\perp$) correlation length was calculated. From the values of $D_\perp$ and of the interlayer distance, the number of stacked layers was estimated to be about 12 for CNT. Reflections in the patterns of the sp^2 adducts samples Figure S4b,c remains at the same 2θ value. MWCNT and MWCNT/PPGP adducts present distances between the structural layers slightly more significant than those of ordered graphite samples (d_{002} = 0.335 nm). In all samples, (112) reflection is negligible.

High-resolution transmission electron microscopy (HR-TEM) was exploited to study the morphology of CNT/PPGP adducts. Various magnifications were adopted. Figure 8 shows HR-TEM micrographs of MWCNT/PPGP$_s$ (A,a), MWCNT/PPGP$_c$ (B,b), SWCNT/PPGP$_s$ (C,c) and SWCNT/PPGPc (D,d) adducts, at lower (A,B,C,D) and at higher magnifications (a,b,c,d).

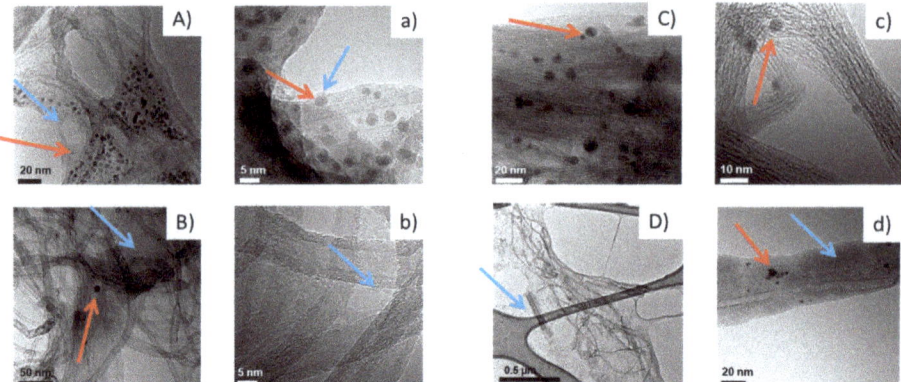

Figure 8. Micrographs of MWCNT/PPGP$_s$ (**A**,**a**), MWCNT/PPGP$_c$ (**B**,**b**), SWCNT/PPGP$_s$ (**C**,**c**) and SWCNT/PPGP$_c$ (**D**,**d**) adducts isolated from 1 mg/mL water dispersions. Micrographs are: low magnification bright-field TEM (**A**–**D**), HR-TEM images (**a**–**d**); (the light blue arrows indicate layers of organic substances; the red arrows indicate spherical organic aggregates).

Micrographs at a lower magnification of each CNT/PPGP adduct in Figure 8 reveal that the length of CNT/PPGP adducts is of the same order of magnitude in samples isolated before and after PPGP treatments. This indicates that the chemical interaction with PPGP and the heating step for the preparation of the covalent adducts do not cause appreciable breaking of the nanotubes.

In the case of MWCNT/PPGP$_s$ and SWCNT/PPGP$_s$ adducts in water dispersions, micrographs at lower magnification in Figure 8A,C reveal carbon aggregates made by pseudo-spherical particles with an average size of about 5–10 nm. Figure 8B,b shows the MWCNT/PPGP$_c$ covalent adduct. It seems that a layer of organic substance (indicated by the light blue arrow) adheres to the carbon allotrope. It appears that the organic substance is probably made by unreacted PPGP, which covers the surface. Also, in the case of MWCNT/PPGP$_c$, a low quantity of spherical organic aggregates (indicated by the red arrow) were detected (average dimension~5–20 nm). It appears that spherical agglomerates are probably made by unreacted PPGP also in this case. It is known from the literature that macromolecules like polyether terminated with cationic or anionic functional groups can generate micelle or more in general supramolecular assembly structures [46].

Micrographs at higher magnification Figure 8a–d allow visualizing walls of nanotubes. In the case of SWCNT, the covalent treatment with PPGP led to a large CNT disentanglement, as shown by the lower number of CNT micrometric bundles and by the presence of individual tubes in a defined space, as observed in many HR-TEM images and represented in Figure 8. The micrograph in Figure 8d shows that the SWCNT skeleton remained intact after the treatment with PPGP oligomer. The CNT surface was thus decorated with PPGP chains, which form condensed polymer layers adhered to the CNT external surface, with a thickness from about 3 to about 10 nm.

Furthermore, a comparison between TEM micrographs at low magnifications of pristine CNT and PPGP adducts are reported in Figure S5. Figure S5a,b show that a high entanglement of nanotubes characterizes pristine MWCNT (a) and SWCNT (b). The functionalization with PPGP improves the disentanglement of CNT, allowing better processability of the carbon allotropes dispersions Figure S5c–f.

3.2. Preparation and Characterization of the Ternary Nano Complex CNT/PPGP/DOX

HR-TEM microscopy Figure 8 allows checking that both supramolecular adducts show on their surface micelle-like structures and adhered polymer chains, and to the contrary, that the covalent adducts show only a regular adherent polymeric layer. In both cases, the possibility of a drug loading seems possible via adsorption on the CNT surface. Previously, it has been shown that the mixing of the SWCNT with DOX leads to the absorption of DOX onto the outer sides of SWCNTs via π–π stacking interactions. It was reported that suitably functionalized SWNT and MWCNT have been found to be non-toxic in mice and can be gradually excreted by the biliary pathway [28]. We explored the possibility of using supramolecular π–π stacking to load a cancer chemotherapy agent DOX on CNT/PPGP adducts for drug delivery applications.

In this work, we describe a previously unreported non-covalent CNT/PPGP/DOX supramolecular nano complex that can be developed for cancer therapy. We have investigated the ability of DOX to interact non-covalently and covalently with CNT functionalized with PPGP.

Figure 9 shows schemes suggesting the structures of both covalent and supramolecular CNT/PPG adducts (Panel A) and the hypothesized ternary nano complex CNT/PPGP/DOX (Panel B).

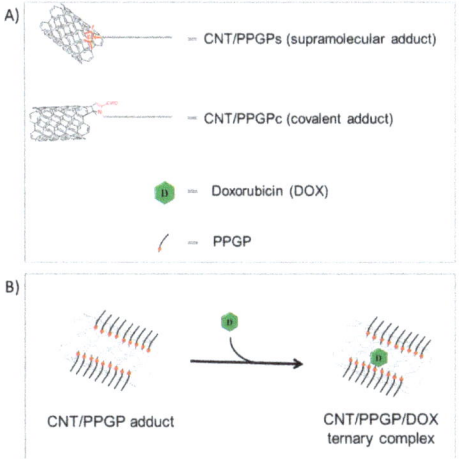

Figure 9. Representation of (**A**) the CNT/PPGP adducts, and (**B**) the hypothesized ternary complex CNT/PPGP/DOX.

The ternary nano complex was prepared, as described in Figure 9B. In brief, DOX hydrochloride was stirred for 16 h at room temperature with the modified nanotubes dispersed in a pH 7.4 PBS buffered solution. The CNT/PPGP/DOX nano complexes were isolated by repeated ultracentrifugation with PBS until the supernatant became colorless. Free, unbound DOX in the CNT supernatant was analyzed by UV-Vis spectroscopy. DOX characteristic absorbance peak at 490 nm was detected Figure S2. The amount of unbound DOX onto the CNT was estimated by measuring the absorbance at 490 nm relative to a calibration curve recorded under the same conditions Figure S2. In Table S1, the amount of loaded DOX is reported for pristine and functionalized CNT. After DOX loading on the PPGP modified carbon nanotubes, two different scenarios were observed for MWCNT and SWCNT respectively: (i) in MWCNT/PPGP/DOX UV-Vis spectrum the absorption band at 490 nm was not detected; (ii) in SWCNT/PPGP/DOX the absorption band at 490 nm is slightly redshifted (Figure S3).

The interaction between DOX and CNT/PPGP adducts was studied by monitoring the emission spectrum of DOX by fluorescence spectrophotometry (Figure 10).

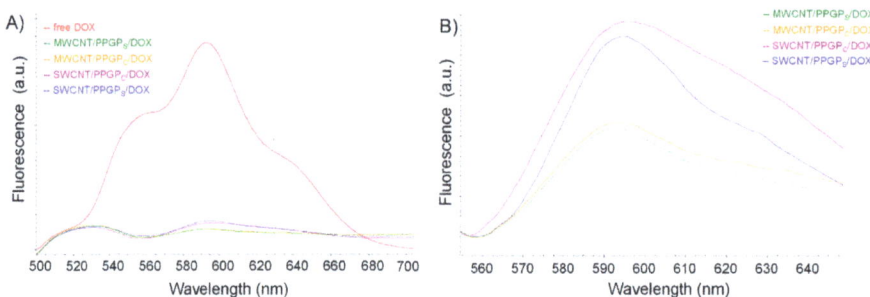

Figure 10. Normalized fluorescence intensities spectra of: (Panel **A**) DOX and CNT/PPGP/DOX nano complexes; (Panel **B**) CNT/PPGP/DOX nano complexes, zoom in the region between 560 to 640 nm. Irradiation at 480 nm.

As can be seen from Figure 10, the fluorescence quenching of DOX was evident for all the nano complexes.

Release profiles of DOX (Figure 11) from all the modified and unmodified CNT at 37 °C were evaluated for up to 72 h at pH 7.4 in PBS (Figure 11A,C), which mimics the acidity of cytoplasm, and at pH 5.5 acetate buffer (Figure 11B,D), which mimics the acid condition of lysosomes, endosomes and cancerous tissues [47]. A slow-release with a pH-sensitive profile for all the investigated samples was observed. Our results were found to be in line with previous studies [48]. At physiological pH, DOX tended to remain bound to the CNT or CNT/PPGP adducts, whereas at acidic pH, the increased protonation of DOX changes both its solubility and hydrophilicity, hence leading to a higher release of the drug from the complexes [49].

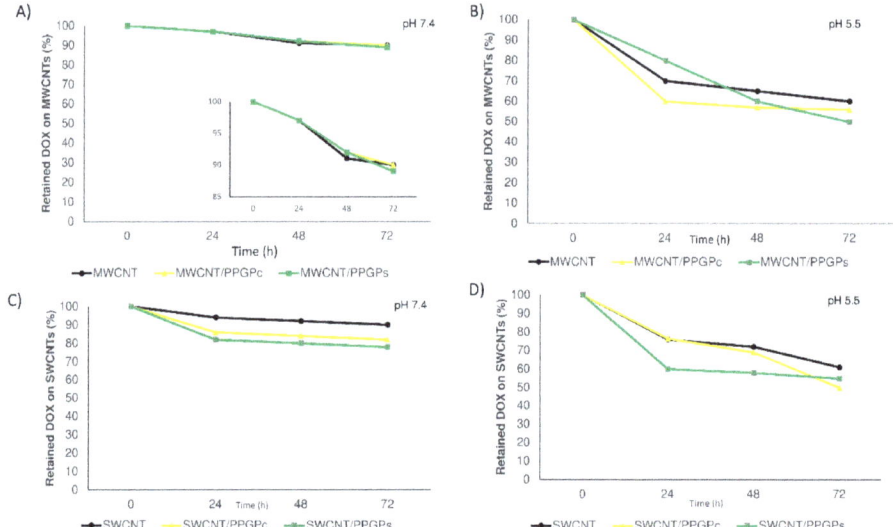

Figure 11. Drug release of the unmodified CNT/DOX and modified CNT/PPGP/DOX nano complexes in PBS buffer at 37 °C at pH 7.4 (**A,C**), MWCNTs and SWCNTs respectively) and in acetate buffer at pH 5.5 (**B,D**), MWCNTs and SWCNTs respectively); magnification in the range from 85% to 100% of retained DOX on MWCNT is also reported in the y-axis (**A**).

No initial burst effect was observed in either of the conditions adopted. After 72 h, at pH 7.4 the amount of released DOX was in the range of 10–11% for MWCNT/DOX and MWCNT/PPGP/DOX complexes Figure 11A and of 10–20% for SWCNT/DOX and SWCNT/PPGP/DOX (Figure 11C). At pH 5.5, a different behavior was observed for CNT/DOX and CNT/PPPG/DOX complexes. Regarding MWCNT/DOX and MWCNT/PPGP/DOX complexes (Figure 11B), the amount of DOX released over a 72-h period was observed to be faster for MWCNT/DOX and MWCNT/PPGP$_c$/DOX, while for MWCNT/PPGP$_s$/DOX a controlled and a linear release was observed. In contrast, for SWCNT/DOX and SWCNT/PPGP/DOX complexes, the release of DOX at pH 5.5 Figure 11D was found to be faster for MWCNT/PPGP$_s$/DOX and lower for MWCNT/DOX and MWCNT/PPGP$_c$/DOX. The different complexes CNT/PPGP/DOX could satisfy different release rates, depending on the type of CNT and the nature of the interaction with PPGP and DOX.

3.3. Cell Viability Assay

In order to estimate the likelihood of our CNT drug delivery systems, the cytotoxicity of MWCNT, MWCNT/PPGP$_s$, MWCNT/PPGP$_c$, MWCNT/DOX, MWCNT/PPGP$_s$/DOX, and MWCNT/PPGP$_c$/DOX on A549 and M14 cell lines was evaluated. A stock dispersion of each CNT was prepared in culture medium at 1 mg/mL and sonicated. This procedure was ineffective for SWCNT, SWCNT/PPGP$_s$, SWCNT/PPGP$_c$, SWCNT/DOX, SWCNT/PPGP$_s$/DOX, and SWCNT/PPGP$_c$/DOX samples because of the persistence of carbon aggregates and agglomerates turned into the absence of a homogenous dispersion suitable for evaluation in cell culture assay. Indeed, the stock dispersion of MWCNT, MWCNT/PPGP$_s$, MWCNT/PPGP$_c$, MWCNT/DOX, MWCNT/PPGP$_s$/DOX, and MWCNT/PPGP$_c$/DOX was further diluted considering the amount of loaded DOX into each sample, as reported in Table S1. Thus, the concentrations were normalized and reported as the DOX amount.

The cytotoxic properties of loaded DOX were compared with that of the free DOX (Figures 12 and 13) at four different concentrations (8.25 µg/mL, 16.5 µg/mL, 33 µg/mL, 66 µg/mL) after 48 h of treatment. Our preliminary results indicated that DOX maintains its inhibitory effect for all the investigated cases (free DOX, covalently loaded DOX, or complexed DOX) on both A549 and M14 cellular lines. Moreover, from our data, it emerges that there is a different behavior for the CNT drug delivery systems with a cell variance. MWCNT/DOX showed lower activity compared to the free DOX on M14 cellular lines; MWCNT/PPGP$_c$/DOX and the free DOX displayed a similar effect on both cellular lines at all the investigated concentration; MWCNT/PPGP$_c$/DOX appeared more efficient than free DOX in A549 cell lines and less efficient than the free DOX in MT14 ones.

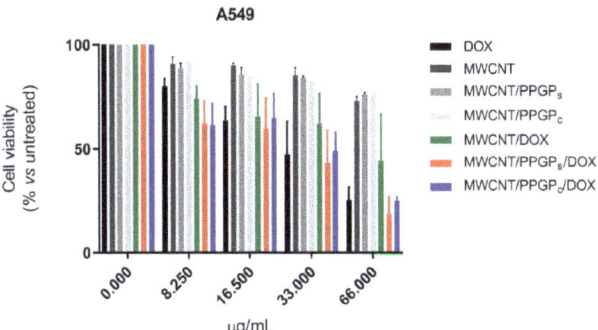

Figure 12. Cell viability (MTT assay) of A549 human melanoma cells untreated and treated for 48 h with different concentrations of free DOX hydrochloride or into both unloaded and loaded CNT. Each point represents mean ± s.e.m. of three separate experiments performed in triplicate.

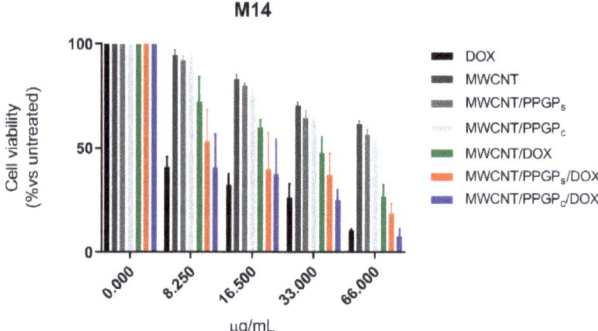

Figure 13. Cell viability (MTT assay) of M14 human lung adenocarcinoma cells untreated and treated for 48 h with different concentrations of free DOX hydrochloride or into both unloaded and loaded CNT. Each point represents mean ± s.e.m. of three separate experiments performed in triplicate.

Notably, the comparable cytotoxicity effects between free DOX and loaded DOX were discussed without considering the delayed release of DOX from the carrier. Considering the amount of DOX released at 48 h (about 35–43% at pH 5.5, Figure 11B, and 8–9% at pH 7.4, Figure 11A), a major cytotoxic effect can be ascribed to the DOX loaded on CNT respected to the free DOX, presumably due to a more efficient internalization route. Further investigation will be devoted to the studies of cellular uptake and intracellular trafficking of our nanocarrier to verify the chance to extend the use of this DDS.

3.4. Computational Studies

Although some molecular dynamics (MD) studies upon DOX/SWCNT systems were already performed [50–55], to the best of our knowledge, this is the first one that takes in consideration even a DOX/MWCNT system and, in particular, the use of opportunely reduced systems which reflects the actual diameter of the CNTs employed for the reported experiments and the CNT/PPGP/DOX ratios obtained from the experimentally observed weight percentages. So, we built four supramolecular systems named SSy1–4 corresponding to the complexes SWCNT/PPGP$_s$/DOX, SWCNT/PPGP$_c$/DOX, and MWCNT/PPGP$_s$/DOX, MWCNT/PPGP$_c$/DOX, respectively, and submitted each of them to 100 ns MD simulations. For each experiment, we used three different systems in which the starting positions of the PPGP and/or DOX molecules were randomly varied.

The studied systems have reached their equilibrium states after about 10 and 15 ns of simulation time for the SSy1,2 and SSy3,4, respectively, as revealed by the root-mean-square displacements (RMSD) of DOX molecules reported, only for SSy2,4, in Figures 14 and 15; it is evident that the fluctuations in RMSDs have reduced significantly after these periods.

In particular, Figure 14 shows the run of SSy2 in which there are two DOX molecules within the CNT at the starting time (0 ns) and 6 DOX molecules at the end time (100 ns). The entrance of the other four molecules, in pairs, takes place at 41.9 and 78.5 ns, respectively, as evidenced by the fluctuations registered in the RMSD graph for the DOX molecules. Interestingly, the six molecules within the CNT cavity were paired to four chloride ions, which coordinates some ammonium groups, some of which were facing each other (Figure 16, left; no sodium ion is present. The remaining DOX molecules are strongly anchored to the external wall of the SWCNT employing π–π interactions and, in some cases, two and even three of them are stacked together; the PPG pendant remains, most of the time, close to the external wall of the CNT (Figure 16, right).

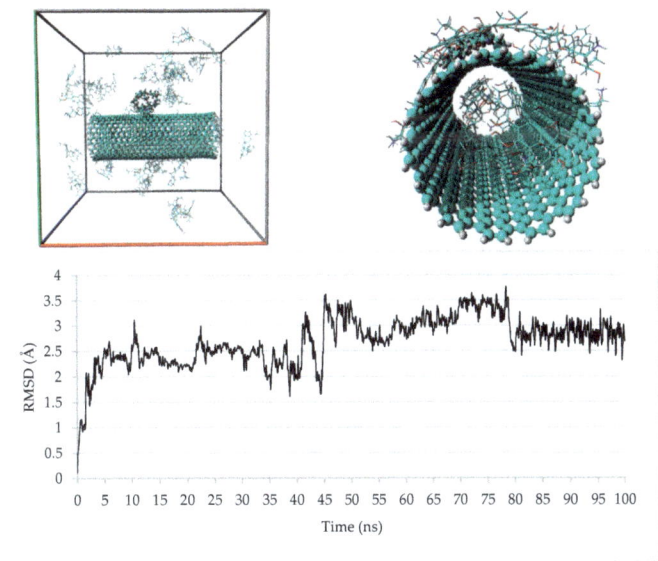

Figure 14. Upper: SSy2 at the start (left) and the end (right) of the MD simulation; bottom: plot of the RMSD of the DOX molecules during the 100 ns of the MD simulation.

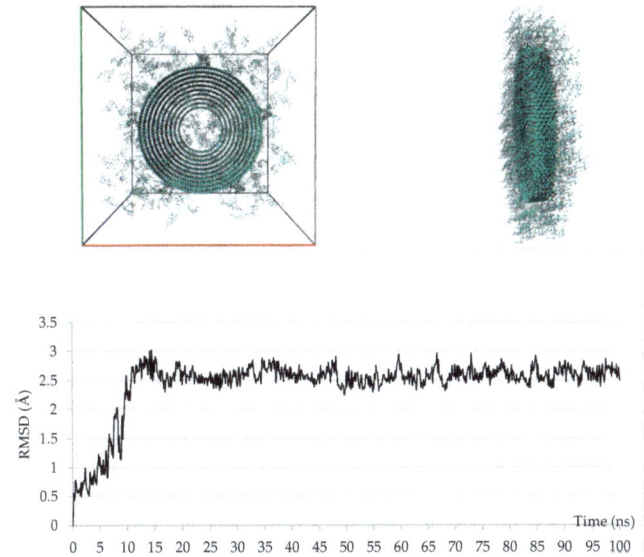

Figure 15. Upper: SSy4 at the start (left) and the end (right) of the MD simulation; bottom: plot of the RMSD of the DOX molecules during the 100 ns of the MD simulation.

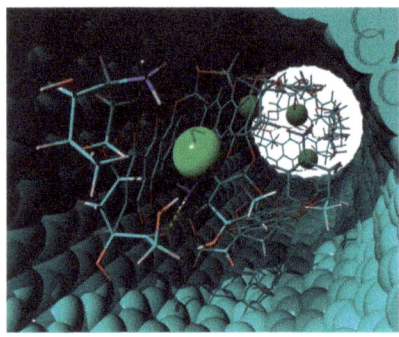 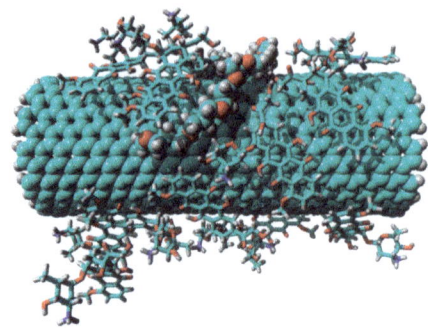

Figure 16. Left: view of the DOX molecules and the chloride ions contained inside the SWCNT cavity of the SSy2 at the end of the 100 ns of the MD simulation; right: view of the DOX molecules and the PPG pendant anchored to the external wall of the SWCNT of the SSy2 at the end of the 100 ns of the MD simulation.

As regards the SSy4 system, at the end of the 100 ns of the MD simulations, there are 8 DOX molecules within the inner CNT cavity, whereas almost all others are mainly adsorbed upon the external CNT surface of the far wall, stacked up to 5 units. Due to the short length of the tube (10 Å), the same DOX molecules are scattered along the two ends, usually with the most extended axis parallel to that of the CNT and stacked on each other.

The two different accommodation of DOX molecules between the two SSy2 and SSy4 systems are in accord to their respective side surface extensions, 1385 Å^2 and 1492 Å^2, except for that occupied by the covalently bound PPGP molecules. Considering that the semi-surface of a DOX molecule corresponds about to 75 Å^2, about 18 upon 45, and 20 upon 354 DOX molecules for the SSy2 and the SSy4 system, respectively, should be able to cover the entire surface of each CNT. This also means that, on the equal surface, the efficiency of the MWCNT is approximately 7.8 times higher than that of the SWCNT. Finally, the ratio of about 1:5 of the PPGP molecules on the equal surface makes the MWCNT much more soluble than the SWCNT.

For the other two systems, SSy1,3, the trend is almost superimposable upon that of the corresponding systems with the covalently linked PPGP moieties. In both cases, the PPGP molecules are found to be sufficiently adherent to the surface of the CNT, mostly thanks to the portion of the PPG moiety; this is in accord with the TEM micrographs C,c and A,a of Figure 8 that highlight the presence of a polymeric layer adhered to the CNT surface. Moreover, for the SSy3 complex, the presence of 12 PPGP molecules strongly adsorbed on its surface make it even more soluble as the SSy4 parent.

4. Conclusions

Nanomedicine and technological nano delivery systems are a rather new but rapidly developing field where molecules in the nanoscale range are employed as diagnostic tools or to deliver therapeutics to specifically targeted sites and with a huge controlled fashion [56–59]. CNT are emerging nanomaterials with massive potential in the diagnostic and therapeutic fields. An effective way to make CNT more biocompatible and to increase their application in medicine is to functionalize the nanotubes. In this paper, we reported the functionalization of single and multi-walled carbon nanotubes with a pyrrole polypropylene glycol derived compound exploiting a *Diels-Alder* reaction. Thermogravimetric analysis and FT-IR spectroscopy showed that the functionalization of the considered adduct with PPGP was successful and that the introduction of a proper amount of modifier was achieved. WAXD shows that the functionalization procedures do not substantially alter per se the bulk structure of carbon nanotubes. The obtained functionalized CNT were then exploited to make a non-covalent CNT/PPGP/DOX supramolecular nano complex. The ability of DOX to interact with the non-covalent and the covalent PPGP modified CNT was investigated by experimental and computational techniques. HR-TEM microscopy confirmed that the covalent adducts show a regular adherent polymeric layer, and the supramolecular adducts contain on their surface both micelle-like structures and adherent polymer chains. MD simulations showed that DOX molecules can be adsorbed to the external wall of the nanotubes or included in their cavity.

Biological studies revealed that the in vitro activity of MWCNT/PPGP$_s$/DOX and MWCNT/PPGP$_c$/DOX are similar to that of the free DOX in A549 and M14 cell lines, although the former activities are actually attributable to a release, at 48 h, of approximately 8% (at pH 7.4) or 40% (at pH 5.5) of DOX.

Moreover, our studies show a different biological behavior between pyrrole functionalized-SWCNT and pyrrole functionalized-MWCNT, although a similar degree of chemical was detected for both materials. The formation of carbon aggregates and agglomerates in biological media for pyrrole functionalized-SWCNT prevented their evaluation, whereas the better dispersibility of pyrrole functionalized-MWCNT allowed the evaluation of their cytotoxicity in cell culture assay.

The use of carbon nanotubes in the drug delivery field seems promising due to the ability of CNT to cross biological barriers. This work paves the way for the facile functionalization of carbon nanotubes exploiting the "pyrrole methodology" for the development of novel technological carbon-based drug delivery systems.

Even if the preliminary biological studies were satisfactory, more mechanistic work is needed to investigate the capabilities of the novel "pyrrole functionalized" CNT to translocate into cells.

Moreover, the intracellular trafficking of MWCNT/PPGP$_s$/DOX, MWCNT/PPGP$_c$/DOX, and released DOX that determines the drug efficacy and the related side effects also need be studied.

Supplementary Materials: The following are available online at http://www.mdpi.com/2079-4991/10/6/1073/s1, WAXD, HSP, and MD simulations details, Figure S1: ^{1}H NMR spectrum (400 MHz, CDCl$_3$) of the O-(2-(2,5-dimethyl-1H-pyrrol-1-yl) propyl)-O'-(2-methoxyethyl)polypropylene glycol (pyrrole polypropylene glycol, PPGP), Figure S2: UV-Vis absorbance spectra of unbonded DOX in PBS supernatant after centrifugation of: (A) MWCNT/DOX, MWCNT/PPGPc/DOX and MWCNT/PPGPs/DOX nano complexes; (B) SWCNT/DOX, SWCNT/PPGPc/DOX and SWCNT/PPGPs/DOX nano complexes. Figure S3: UV-Vis absorbance spectra of PBS solutions of free DOX (pink), MWCNT/PPGP/DOX covalent (yellow), and supramolecular (black) adducts, SWCNT/PPGP/DOX covalent (purple) and supramolecular (light blue) adducts. Figure S4: WAXD patterns of MWCNT (a), MWCNT/PPGP$_s$ (b), and MWCNT/PPGP$_c$ (c), Figure S5: TEM micrographs at low magnifications of pristine CNT and PPGP adducts, Table S1: Drug loading of the pristine and modified CNT, Table S2: Results of the inspections performed on the 1 mg/mL dispersions of the reported sp^2 carbon allotropes (CA) in the listed solvents. The label 'good' indicates the stability of the dispersion, while the label 'bad' indicates a separation between the CA and the solvent. Section S1: Wide-Angle X-ray Diffraction Details. Section S2: Hansen Solubility Parameters (HSP) Details. Section S3: Molecular Dynamics Details.

Author Contributions: Conceptualization, A.R. and V.B.; methodology, C.P.; G.F.; A.C.E.G., and V.C.; validation, L.R.; G.F.; A.C.E.G. and V.C.; formal analysis, A.R.; V.C. and V.B.; synthesis and characterization, C.P.; L.R., and V.B.; biological investigation, A.C.E.G., and V.C.; computational investigation, G.F. and A.R.; data curation, C.P.; L.R., G.F.; V.C.; M.G.; A.R., and V.B.; writing—original draft preparation, C.P.; G.F.; A.C.E.G.; V.C.; A.R. and V.B.; writing—review and editing, C.P.; G.F.; A.C.E.G.; V.C.; L.R.; M.G.; A.R. and V.B.; supervision, A.R. and V.B.; project administration, A.R. and V.B. All authors have read and agreed to the published version of the manuscript.

Funding: This research received no external funding.

Acknowledgments: Authors gratefully acknowledge Andrea Serafini (Politecnico di Milano) for HR-TEM micrographs and Luigi Brambilla (Politecnico di Milano) for Fluorescence spectrophotometry.

Conflicts of Interest: The authors declare no conflict of interest.

References

1. Kroto, H.W.; Mckay, K. The Formation of Quasi-Icosahedral Spiral Shell Carbon Particles. *Nature* **1988**, *331*, 328–331. [CrossRef]
2. Terrones, M.; Botello-Mendez, A.R.; Campos-Delgado, J.; Lopez-Urias, F.; Vega-Cantu, Y.I.; Rodriguez-Macias, F.J.; Elias, A.L.; Munoz-Sandoval, E.; Cano-Marquez, A.G.; Charlier, J.C.; et al. Graphene and graphite nanoribbons: Morphology, properties, synthesis, defects and applications. *Nano Today* **2010**, *5*, 351–372. [CrossRef]

3. Bethune, D.S.; Kiang, C.H.; Devries, M.S.; Gorman, G.; Savoy, R.; Vazquez, J.; Beyers, R. Cobalt-Catalyzed Growth of Carbon Nanotubes with Single-Atomic-Layer walls. *Nature* **1993**, *363*, 605–607. [CrossRef]
4. Iijima, S.; Ichihashi, T. Single-Shell Carbon Nanotubes of 1-Nm Diameter. *Nature* **1993**, *363*, 603–605. [CrossRef]
5. Iijima, S. Helical Microtubules of Graphitic Carbon. *Nature* **1991**, *354*, 56–58. [CrossRef]
6. Monthioux, M.; Kuznetsov, V.L. Who should be given the credit for the discovery of carbon nanotubes? *Carbon* **2006**, *44*, 1621–1623. [CrossRef]
7. Novoselov, K.S.; Geim, A.K.; Morozov, S.V.; Jiang, D.; Zhang, Y.; Dubonos, S.V.; Grigorieva, I.V.; Firsov, A.A. Electric field effect in atomically thin carbon films. *Science* **2004**, *306*, 666–669. [CrossRef]
8. Geim, A.K.; Novoselov, K.S. The rise of graphene. *Nat. Mater.* **2007**, *6*, 183–191. [CrossRef]
9. Allen, M.J.; Tung, V.C.; Kaner, R.B. Honeycomb Carbon: A Review of Graphene. *Chem. Rev.* **2010**, *110*, 132–145. [CrossRef]
10. Zhu, Y.W.; Murali, S.; Cai, W.W.; Li, X.S.; Suk, J.W.; Potts, J.R.; Ruoff, R.S. Graphene and Graphene Oxide: Synthesis, Properties, and Applications. *Adv. Mater.* **2010**, *22*, 3906–3924. [CrossRef]
11. Geng, Y.; Wang, S.J.; Kim, J.K. Preparation of graphite nanoplatelets and graphene sheets. *J. Colloid Interface Sci.* **2009**, *336*, 592–598. [CrossRef] [PubMed]
12. Kavan, L.; Yum, J.H.; Gratzel, M. Optically Transparent Cathode for Dye-Sensitized Solar Cells Based on Graphene Nanoplatelets. *ACS Nano* **2011**, *5*, 165–172. [CrossRef] [PubMed]
13. Nieto, A.; Lahiri, D.; Agarwal, A. Synthesis and properties of bulk graphene nanoplatelets consolidated by spark plasma sintering. *Carbon* **2012**, *50*, 4068–4077. [CrossRef]
14. Paton, K.R.; Varrla, E.; Backes, C.; Smith, R.J.; Khan, U.; O'Neill, A.; Boland, C.; Lotya, M.; Istrate, O.M.; King, P.; et al. Scalable production of large quantities of defect-free few-layer graphene by shear exfoliation in liquids. *Nat. Mater.* **2014**, *13*, 624–630. [CrossRef]
15. Rao, C.N.R.; Satishkumar, B.C.; Govindaraj, A.; Nath, M. Nanotubes. *ChemPhysChem* **2001**, *2*, 78–105. [CrossRef]
16. De Heer, W.A. Nanotubes and the pursuit of applications. *MRS Bull.* **2004**, *29*, 281–285. [CrossRef]
17. Uchida, T.; Kumar, S. Single wall carbon nanotube dispersion and exfoliation in polymers. *J. Appl. Polym. Sci.* **2005**, *98*, 985–989. [CrossRef]
18. Jorio, A.; Dresselhaus, G.; Dresselhaus, M.S. *Carbon Nanotubes: Advanced Topics in the Synthesis, Structure, Properties and Applications.* [In: Top. Appl. Phys., 2008; 111]; Springer GmbH: Nettetal, Germany, 2008; p. 720.
19. Kasumov, A.Y.; Deblock, R.; Kociak, M.; Reulet, B.; Bouchiat, H.; Khodos, I.I.; Gorbatov, Y.B.; Volkov, V.T.; Journet, C.; Burghard, M. Supercurrents through single-walled carbon nanotubes. *Science* **1999**, *284*, 1508–1511. [CrossRef]
20. Niu, C.M.; Sichel, E.K.; Hoch, R.; Moy, D.; Tennent, H. High power electrochemical capacitors based on carbon nanotube electrodes. *Appl. Phys. Lett.* **1997**, *70*, 1480–1482. [CrossRef]
21. Baughman, R.H.; Cui, C.X.; Zakhidov, A.A.; Iqbal, Z.; Barisci, J.N.; Spinks, G.M.; Wallace, G.G.; Mazzoldi, A.; De Rossi, D.; Rinzler, A.G.; et al. Carbon nanotube actuators. *Science* **1999**, *284*, 1340–1344. [CrossRef]
22. Ago, H.; Petritsch, K.; Shaffer, M.S.P.; Windle, A.H.; Friend, R.H. Composites of carbon nanotubes and conjugated polymers for photovoltaic devices. *Adv. Mater.* **1999**, *11*, 1281–1285. [CrossRef]
23. Bachtold, A.; Hadley, P.; Nakanishi, T.; Dekker, C. Logic circuits with carbon nanotube transistors. *Science* **2001**, *294*, 1317–1320. [CrossRef] [PubMed]
24. Ajayan, P.M.; Iijima, S. Capillarity-Induced Filling of Carbon Nanotubes. *Nature* **1993**, *361*, 333–334. [CrossRef]
25. Andrews, R.; Weisenberger, M.C. Carbon nanotube polymer composites. *Curr. Opin. Solid State Mater. Sci.* **2004**, *8*, 31–37. [CrossRef]
26. Xie, X.L.; Mai, Y.W.; Zhou, X.P. Dispersion and alignment of carbon nanotubes in polymer matrix: A review. *Mater. Sci. Eng. R-Rep.* **2005**, *49*, 89–112. [CrossRef]
27. Moniruzzaman, M.; Winey, K.I. Polymer nanocomposites containing carbon nanotubes. *Macromolecules* **2006**, *39*, 5194–5205. [CrossRef]
28. Kostarelos, K.; Lacerda, L.; Pastorin, G.; Wu, W.; Wieckowski, S.; Luangsivilay, J.; Godefroy, S.; Pantarotto, D.; Briand, J.P.; Muller, S.; et al. Cellular uptake of functionalized carbon nanotubes is independent of functional group and cell type. *Nat. Nanotechnol.* **2007**, *2*, 108–113. [CrossRef]

29. Pantarotto, D.; Singh, R.; McCarthy, D.; Erhardt, M.; Briand, J.P.; Prato, M.; Kostarelos, K.; Bianco, A. Functionalized carbon nanotubes for plasmid DNA gene delivery. *Angew. Chem.-Int. Ed.* **2004**, *43*, 5242–5246. [CrossRef]
30. Jiang, B.P.; Zhou, B.; Lin, Z.X.; Liang, H.; Shen, X.C. Recent Advances in Carbon Nanomaterials for Cancer Phototherapy. *Chem. Eur. J.* **2019**, *25*, 3993–4004. [CrossRef]
31. Yang, M.; Zhang, M.F. Biodegradation of Carbon Nanotubes by Macrophages. *Front. Mater.* **2019**, *6*, 225. [CrossRef]
32. Singh, P.; Campidelli, S.; Giordani, S.; Bonifazi, D.; Bianco, A.; Prato, M. Organic functionalisation and characterisation of single-walled carbon nanotubes. *Chem. Soc. Rev.* **2009**, *38*, 2214–2230. [CrossRef] [PubMed]
33. Barbera, V.; Brambilla, L.; Porta, A.; Bongiovanni, R.; Vitale, A.; Torrisi, G.; Galimberti, M. Selective edge functionalization of graphene layers with oxygenated groups by means of Reimer-Tiemann and domino Reimer-Tiemann/Cannizzaro reactions. *J. Mater. Chem. A* **2018**, *6*, 7749–7761. [CrossRef]
34. Galimberti, M.; Barbera, V.; Citterio, A.; Sebastiano, R.; Truscello, A.; Valerio, A.M.; Conzatti, L.; Mendichi, R. Supramolecular interactions of carbon nanotubes with biosourced polyurethanes from 2-(2,5-dimethyl-1H-pyrrol-1-yl)-1,3-propanediol. *Polymer* **2015**, *63*, 62–70. [CrossRef]
35. Barbera, V.; Guerra, S.; Brambilla, L.; Maggio, M.; Serafini, A.; Conzatti, L.; Vitale, A.; Galirnberti, M. Carbon Papers and Aerogels Based on Graphene Layers and Chitosan: Direct Preparation from High Surface Area Graphite. *Biomacromolecules* **2017**, *18*, 3978–3991. [CrossRef]
36. Anastas, P.; Eghbali, N. Green Chemistry: Principles and Practice. *Chem. Soc. Rev.* **2010**, *39*, 301–312. [CrossRef]
37. Galimberti, M.; Barbera, V.; Guerra, S.; Bernardi, A. Facile Functionalization of Sp(2) Carbon Allotropes with a Biobased Janus Molecule. *Rubber Chem. Technol.* **2017**, *90*, 285–307. [CrossRef]
38. Barbera, V.; Brambilla, L.; Milani, A.; Palazzolo, A.; Castiglioni, C.; Vitale, A.; Bongiovanni, R.; Galimberti, M. Domino Reaction for the Sustainable Functionalization of Few-Layer Graphene. *Nanomaterials* **2019**, *9*, 44. [CrossRef]
39. Barbera, V.; Bernardi, A.; Palazzolo, A.; Rosengart, A.; Brambilla, L.; Galimberti, M. Facile and sustainable functionalization of graphene layers with pyrrole compounds. *Pure Appl. Chem.* **2018**, *90*, 253–270. [CrossRef]
40. Russo, A.; Cardile, V.; Graziano, A.C.E.; Formisano, C.; Rigano, D.; Canzoneri, M.; Bruno, M.; Senatore, F. Comparison of essential oil components and in vitro anticancer activity in wild and cultivated Salvia verbenaca. *Nat. Prod. Res.* **2015**, *29*, 1630–1640. [CrossRef]
41. Rescifina, A.; Surdo, E.; Cardile, V.; Avola, R.; Eleonora Graziano, A.C.; Stancanelli, R.; Tommasini, S.; Pistarà, V.; Ventura, C.A. Gemcitabine anticancer activity enhancement by water soluble celecoxib/sulfobutyl ether-beta-cyclodextrin inclusion complex. *Carbohydr. Polym.* **2019**, *206*, 792–800. [CrossRef]
42. Jain, A.K.; Swarnakar, N.K.; Das, M.; Godugu, C.; Singh, R.P.; Rao, P.R.; Jain, S. Augmented Anticancer Efficacy of Doxorubicin-Loaded Polymeric Nanoparticles after Oral Administration in a Breast Cancer Induced Animal Model. *Mol. Pharm.* **2011**, *8*, 1140–1151. [CrossRef] [PubMed]
43. Nanotube Modeler: Generation of Nano-Geometries. Available online: http://www.jcrystal.com/products/wincnt/ (accessed on 9 May 2020).
44. The DrugBank Database. Available online: https://www.drugbank.ca/ (accessed on 9 May 2020).
45. Duan, Y.; Wu, C.; Chowdhury, S.; Lee, M.C.; Xiong, G.; Zhang, W.; Yang, R.; Cieplak, P.; Luo, R.; Lee, T.; et al. A point-charge force field for molecular mechanics simulations of proteins based on condensed-phase quantum mechanical calculations. *J. Comput. Chem.* **2003**, *24*, 1999–2012. [CrossRef] [PubMed]
46. Kim, J.O.; Kabanov, A.V.; Bronich, T.K. Polymer micelles with cross-linked polyanion core for delivery of a cationic drug doxorubicin. *J. Control. Release* **2009**, *138*, 197–204. [CrossRef] [PubMed]
47. Mindell, J.A. Lysosomal Acidification Mechanisms. *Annu. Rev. Physiol.* **2012**, *74*, 69–86. [CrossRef]
48. Zhang, X.K.; Meng, L.J.; Lu, Q.H.; Fei, Z.F.; Dyson, P.J. Targeted delivery and controlled release of doxorubicin to cancer cells using modified single wall carbon nanotubes. *Biomaterials* **2009**, *30*, 6041–6047. [CrossRef]
49. Liu, Z.; Sun, X.M.; Nakayama-Ratchford, N.; Dai, H.J. Supramolecular chemistry on water-soluble carbon nanotubes for drug loading and delivery. *ACS Nano* **2007**, *1*, 50–56. [CrossRef]
50. Sornmee, P.; Rungrotmongkol, T.; Saengsawang, O.; Arsawang, U.; Remsungnen, T.; Hannongbua, S. Understanding the Molecular Properties of Doxorubicin Filling Inside and Wrapping Outside Single-Walled Carbon Nanotubes. *J. Comput. Theor. Nanosci.* **2011**, *8*, 1385–1391. [CrossRef]

51. Izadyar, A.; Farhadian, N.; Chenarani, N. Molecular dynamics simulation of doxorubicin adsorption on a bundle of functionalized CNT. *J. Biomol. Struct. Dyn.* **2016**, *34*, 1797–1805. [CrossRef]
52. Sadaf, S.; Walder, L. Doxorubicin Adsorbed on Carbon Nanotubes: Helical Structure and New Release Trigger. *Adv. Mater. Interfaces* **2017**, *4*, 1700649. [CrossRef]
53. Zhang, L.; Peng, G.T.; Li, J.C.; Liang, L.J.; Kong, Z.; Wang, H.B.; Jia, L.J.; Wang, X.P.; Zhang, W.; Shen, J.W. Molecular dynamics study on the configuration and arrangement of doxorubicin in carbon nanotubes. *J. Mol. Liq.* **2018**, *262*, 295–301. [CrossRef]
54. Contreras, M.L.; Torres, C.; Villarroel, I.; Rozas, R. Molecular dynamics assessment of doxorubicin-carbon nanotubes molecular interactions for the design of drug delivery systems. *Struct. Chem.* **2019**, *30*, 369–384. [CrossRef]
55. Kordzadeh, A.; Amjad-Iranagh, S.; Zarif, M.; Modarress, H. Adsorption and encapsulation of the drug doxorubicin on covalent functionalized carbon nanotubes: A scrutinized study by using molecular dynamics simulation and quantum mechanics calculation. *J. Mol. Graph. Modell.* **2019**, *88*, 11–22. [CrossRef] [PubMed]
56. Venuti, V.; Crupi, V.; Fazio, B.; Majolino, D.; Acri, G.; Testagrossa, B.; Stancanelli, R.; De Gaetano, F.; Gagliardi, A.; Paolino, D.; et al. Physicochemical Characterization and Antioxidant Activity Evaluation of Idebenone/Hydroxypropyl-beta-Cyclodextrin Inclusion Complex dagger. *Biomolecules* **2019**, *9*, 531. [CrossRef]
57. Floresta, G.; Rescifina, A. Metyrapone-β-cyclodextrin supramolecular interactions inferred by complementary spectroscopic/spectrometric and computational studies. *J. Mol. Struct.* **2019**, *1176*, 815–824. [CrossRef]
58. Floresta, G.; Punzo, F.; Rescifina, A. Supramolecular host-guest interactions of pseudoginsenoside F11 with β-and γ-cyclodextrin: Spectroscopic/spectrometric and computational studies. *J. Mol. Struct.* **2019**, *1195*, 387–394. [CrossRef]
59. Greish, K.; Salerno, L.; Al Zahrani, R.; Amata, E.; Modica, M.; Romeo, G.; Marrazzo, A.; Prezzavento, O.; Sorrenti, V.; Rescifina, A. Novel structural insight into inhibitors of heme oxygenase-1 (HO-1) by new imidazole-based compounds: Biochemical and in vitro anticancer activity evaluation. *Molecules* **2018**, *23*, 1209. [CrossRef] [PubMed]

© 2020 by the authors. Licensee MDPI, Basel, Switzerland. This article is an open access article distributed under the terms and conditions of the Creative Commons Attribution (CC BY) license (http://creativecommons.org/licenses/by/4.0/).

Article

Toxicity of Carbon Nanomaterials and Their Potential Application as Drug Delivery Systems: In Vitro Studies in Caco-2 and MCF-7 Cell Lines

Rosa Garriga [1,*], Tania Herrero-Continente [2], Miguel Palos [3], Vicente L. Cebolla [4], Jesús Osada [2,5], Edgar Muñoz [4] and María Jesús Rodríguez-Yoldi [3,5,*]

1. Departamento de Química Física, Universidad de Zaragoza, 50009 Zaragoza, Spain
2. Departamento de Bioquímica y Biología Molecular, Universidad de Zaragoza, 50013 Zaragoza, Spain; taniaherrero1992@gmail.com (T.H.-C.); josada@unizar.es (J.O.)
3. Departamento de Farmacología y Fisiología, Universidad de Zaragoza, 50013 Zaragoza, Spain; mpalosmarg@gmail.com
4. Instituto de Carboquímica ICB-CSIC, Miguel Luesma Castán 4, 50018 Zaragoza, Spain; vcebolla@icb.csic.es (V.L.C.); edgar@icb.csic.es (E.M.)
5. CIBEROBN (ISCIII), IIS Aragón, IA2, 50009 Zaragoza, Spain
* Correspondence: rosa@unizar.es (R.G.); mjrodyol@unizar.es (M.J.R.-Y.); Tel.: +34-976-762294 (R.G.); +34-976-761649 (M.J.R-Y.)

Received: 23 July 2020; Accepted: 16 August 2020; Published: 18 August 2020

Abstract: Carbon nanomaterials have attracted increasing attention in biomedicine recently to be used as drug nanocarriers suitable for medical treatments, due to their large surface area, high cellular internalization and preferential tumor accumulation, that enable these nanomaterials to transport chemotherapeutic agents preferentially to tumor sites, thereby reducing drug toxic side effects. However, there are widespread concerns on the inherent cytotoxicity of carbon nanomaterials, which remains controversial to this day, with studies demonstrating conflicting results. We investigated here in vitro toxicity of various carbon nanomaterials in human epithelial colorectal adenocarcinoma (Caco-2) cells and human breast adenocarcinoma (MCF-7) cells. Carbon nanohorns (CNH), carbon nanotubes (CNT), carbon nanoplatelets (CNP), graphene oxide (GO), reduced graphene oxide (GO) and nanodiamonds (ND) were systematically compared, using Pluronic F-127 dispersant. Cell viability after carbon nanomaterial treatment followed the order CNP < CNH < RGO < CNT < GO < ND, being the effect more pronounced on the more rapidly dividing Caco-2 cells. CNP produced remarkably high reactive oxygen species (ROS) levels. Furthermore, the potential of these materials as nanocarriers in the field of drug delivery of doxorubicin and camptothecin anticancer drugs was also compared. In all cases the carbon nanomaterial/drug complexes resulted in improved anticancer activity compared to that of the free drug, being the efficiency largely dependent of the carbon nanomaterial hydrophobicity and surface chemistry. These fundamental studies are of paramount importance as screening and risk-to-benefit assessment towards the development of smart carbon nanomaterial-based nanocarriers.

Keywords: cytotoxicity; carbon nanomaterials; drug delivery; doxorubicin; camptothecin; Caco-2; MCF-7

---

1. Introduction

Carbon nanomaterials are promising new materials to be used as drug nanocarriers suitable for medical treatments in biomedicine, due to their large surface area and chemical stability that allows efficient loading of drugs via both covalent and non-covalent interactions [1–3]. Although their interaction with lipid membranes and their subsequent intracellular transport is poorly understood, it has been demonstrated that they can enter cells using various endocytic processes [4,5]. A combination of increased tumor vascular permeability and insufficient lymphatic drainage, resulting in what is termed as enhanced permeability and retention (EPR) effect, enables these nanoparticles to transport chemotherapeutic agents preferentially to tumor sites as compared to healthy tissues, thereby reducing toxic side effects [6]. Furthermore, these systems could be used for formulation of hydrophobic molecules which lack of suitable physicochemical characteristics required for development of stable pharmaceutical dosage form. Transition metal contamination from synthesis procedures can be avoided by purification [7,8]. Poor aqueous dispersibility and high aggregation tendency of pristine carbon nanomaterials can be sorted out by appropriate surface functionalization towards their applications as nanocarriers [9]. Functionalization of carbon nanomaterials can be achieved by either non-covalently coatings with amphiphilic macromolecules like lipid, polymers and surfactants, or covalently with hydrophilic functional groups. Specifically, in vivo studies have shown that functionalization with polyethylene glycol (PEG) allows to achieve prolonged circulation half-life, resulting in so-called 'stealth' behavior, and therefore improved accumulation in tumor by escaping opsonization-induced reticuloendothelial system (RES) clearance [10–13], making these nanocarriers good candidates for cancer diagnostics and treatment. Also PEGylated nanomaterials exhibited remarkably reduced in vivo toxicity, avoiding accumulation in liver or spleen.

To further enhance the therapeutic efficacy of drugs and simultaneously diminish their undesirable systemic side effects, different small targeting molecules such as folic acid (FA) [14,15], ligands with strong affinity against a given receptor overexpressed in a tumor [16], antibodies that recognize tumor-associated antigens [17–20] and also magnetic nanoparticles [3,15] can be further incorporated onto the drug-loaded carbon nanomaterials to confer either active targeting capabilities via receptor-mediated endocytosis or local nanocarrier accumulation induced by external magnetic field. However, addition of excess targeting ligands also increases clearance by the RES because more proteins are now 'visible' on the surface than with PEG. Also, adsorption of proteins and other biomolecules could shield the targeting ligands that have been grafted onto the surface of nanocarriers from binding to their receptors on tumors. 'Stimuli-responsive' drug delivery, that is, strategies to release drug cargo upon experiencing certain tumor-specific triggers (i.e., higher temperature and lower pH) can be extremely useful for selective and controllable drug release [15,21].

There is a trend to combine the therapeutics and early diagnostics (namely theranostics) together [3,18,22–26]. By combining imaging labels with therapeutics in the same platform, the location of the tumor can be precisely delineated, and the optimal drug doses as well as therapeutic time frame could be determined by acquiring the real-time drug distribution information in vivo. Imaging tags like radioactive nuclides [27–30] and fluorescence probes [30,31] can be conjugated in carbon multifunctional nanoplatforms to observe their intracellular trafficking and biodistribution in vitro and in vivo. Raman signals from nanocarbon materials can also provide a reliable method to monitor their distribution and metabolism in vivo [32,33].

In addition, carbon nanotubes, carbon nanohorns and reduced graphene demonstrate strong optical absorption in the near-infrared (NIR) region, making them promising materials for use in the photothermal ablation of tumors [25,34–36]. Carbon nanomaterials also offer promise in combination therapy, that generally refers to two or more therapeutic agents co-delivered simultaneously, and is becoming more popular because it generates synergistic anticancer effects, enabling a low dosage of each compound and overcoming multi-drug resistance (MDR) cancer [37,38]. While the delivery of cancer chemotherapeutic agents with carbon nanomaterials has been more widely attempted, carbon nanomaterials have also demonstrated promising potentials to be used to deliver many non-anticancer drugs, such as antimicrobial, anti-inflammatory, antihypertensive and anti-oxidant agents [1,26].

However, the biomedical applications of carbon nanomaterials arouse serious concerns, as more information on the pharmacokinetics, metabolism, long-term fate and toxicity is essential [27,29,31,39–41]. The issues of toxicity surrounding the biomedical applications of carbon nanomaterials still remain controversial to this date, with studies demonstrating conflicting results [42–45]. Carbon nanomaterial-based drug delivery systems are still considered far from being accepted for use in actual clinical settings. Progress towards clinical trials will depend on the outcomes of efficacy and toxicology studies, which will provide the necessary risk-to-benefit assessments for carbon-based materials. Fundamental studies regarding the impact of size, shape, aggregation degree and functional groups of carbon nanomaterials are needed to provide the design criteria for successful nanomaterial-based strategies. In carbon nanomaterial safety assessment, in vitro cytotoxicity tests are an important research subject because they are fast, reproducible and easy to control the consistency of experimental conditions and should complement and/or supplement in vivo-animal tests. In vitro studies are able to provide information on the biological fate of nanomaterials at the cellular or multicellular levels. In the literature, there is a lack of comparative studies, as extensive variations in the nanomaterial source, functionalization, and experimental conditions do not allow direct comparison of the different results.

We here investigated in vitro toxicity of various Pluronic (F-127)-dispersed carbon nanomaterials in human epithelial colorectal adenocarcinoma (Caco-2) and human breast adenocarcinoma (MCF-7) cell lines. Representative examples, with different size, shape and functional groups on the surface, such as carbon nanohorns (CNH), carbon nanotubes (CNT), graphene nanoplatelets (CNP), graphene oxide (GO), reduced graphene oxide (RGO) and carbon nanodiamonds (ND) were systematically compared under the same experimental conditions.

Furthermore, their potential in the field of drug delivery of anticancer drugs, such as doxorubicin (DOX) and camptothecin (CPT), was also compared in this work. DOX and CPT have been chosen here as examples of hydrophilic and hydrophobic drugs, respectively. DOX belongs to anthracyclines, topoisomerase II inhibitors exhibiting multiple mechanisms of action and high clinical effectiveness against many types of cancer. Notably, cardiotoxicity is a major concern during therapy as it may be dose-limiting [46]. More effective and safer ways of delivering anthracyclines are hence of significant research interest. Also, resistance to anthracyclines and other chemotherapeutics due to P-glycoprotein (P-gp), a membrane transporter that actively pumps doxorubicin out of the cell, is a frequent problem in cancer treatment [47]. It has been reported that nanocarriers enhance doxorubicin uptake in drug-resistant cancer cells [48–50]. Thus, DOX enters the cells attached to nanocomposites bypassing the P-gp transporter, detaches from the nanocomposite surface following natural acidification of endosomes, and migrates reaching the cell nucleus. On the other hand, CPT is a potent anticancer agent with topoisomerase I-inhibiting activity. However, its practical use in viable cancer therapeutic systems is greatly hampered due to its low solubility in aqueous media [51]. The need to formulate water-soluble salts of CPT (that is, alkaline solutions for intravenous injections) led to chemical modifications of the molecule with loss of anti-tumor activity [52,53]. Thus, developing new drug delivery nanocarriers for CPT able to transport and deliver the drug inside the cancer cells has recently received considerable attention [54].

Finally, it has to be noted that the cancer cell lines targeted here, Caco-2 and MCF-7, correspond to cancers among those having the highest incidence in Western countries, hence the interest of anticancer therapeutic studies performed on them.

2. Materials and Methods

2.1. Carbon Nanomaterials

CNT used here are short (average length < 1 μm) and purified (95% wt. %) multi-walled carbon nanotubes from Nanocyl S.A. (Sambreville, Belgium), NANOCYL® NC3150™, produced via catalytic chemical vapor deposition (CCVD) process. CNH were single wall carbon nanohorns provided by Carbonium Srl (Padua, Italy), produced without any catalyst, by rapid condensation of small carbon clusters (C_2 and C_3) resulting from direct vaporization of graphite [55]). Single layer graphene oxide (GO, purity 99 wt. %) was supplied by Cheap Tubes Inc. (Grafton, VT, USA). RGO was from

Sigma-Aldrich (777684, Darmstadt, Germany). CNP (purity 91 at.%.) and detonation nanodiamonds (ND, purified/grade G01) were purchased from PlasmaChem GmbH (Berlin, Germany).

2.2. Characterization of Carbon Nanomaterials

Transmission electron microscopes (TEM, Tecnai T20 and Tecnai F30, FEI, Hillsboro, OR, USA, operating at 200 and 300 KV, respectively) were used to characterize the structural features of carbon nanomaterials. During sample preparation, nanomaterials were dispersed in ethanol, and a drop was placed onto carbon coated copper grids, the sample excess was wicked away by means of a Kimwipe and allowed to dry under ambient conditions. Prior to TEM imaging, the samples on the grids were placed in a O_2-Ar (20% O_2) plasma cleaner (Model 1020 Fischione, Hanau, Germany) for 5–10 s to remove organic (hydrocarbon) contamination.

X-ray photoelectron spectroscopy (XPS) was performed on powder samples deposited onto double-sided carbon tape using an ESCA Plus spectrometer (Omicron, Taunusstein, Germany) provided with a Mg anode (1253.6 eV) working at 225 W (15 mA, 15 kV). CasaXPS software (version 2.3.15, accessed on 1 June 2020) was used for the peak deconvolution and Shirley type baseline correction was applied.

Nitrogen adsorption-desorption isotherms were measured at 77 K (Micromeritics ASAP 2020, Micromeritics Instrument Corp., Norcross, GA, USA) and surface area measurements of the powder samples were obtained using the Brunauer–Emmett–Teller (BET) method at values of relative pressure (p/p_0) between 0.05 and 0.3.

2.3. Dispersion of Carbon Nanomaterials

Pluronic® F-127 (F-127), suitable for cell culture, average molecular weight 12.6 μDa, was purchased from Sigma.Aldrich (Darmstadt, Germany). F-127 solutions at 15 μg·mL^{-1} and 10 min bath sonication (100 W Branson 2510 bath sonicator, Branson Ultrasonics, Danbury, CT, USA) were used here to assist in carbon nanomaterials dispersion at 3.0 and 0.6 μg·mL^{-1} concentration in cell culture medium.

Dispersions of carbon nanomaterials were prepared in cell culture medium without fetal bovine serum (FBS), as it is known that bovine serum albumin (BSA) has different affinity towards carbon nanomaterials. Thus, it has been reported that BSA readily adsorbed on GO, resulting in a decrease in GO toxicity. In contrast, BSA loading capacity was ~9-fold lower for MWCNT [56].

DOX and CPT loading on carbon nanomaterials was performed by simply mixing of solutions in cell culture media, agitated by using a vortex mixer and kept overnight in dark at room temperature. Due to its poor solubility in aqueous media, CPT was initially dissolved in dimethyl sulfoxide (DMSO, ≥99.9%, from Sigma-Aldrich, Darmstadt, Germany) to a concentration of 1.6 mg·mL^{-1}, and then diluted on cell culture medium to the required working concentrations.

2.4. Cell Lines and Cell Culture

Human Caco-2 cell line (TC7 clone) was kindly provided by Edith Brot-Laroche (Université Pierre et Marie Curie-Paris 6, UMR S 872, Les Cordeliers, France). Caco-2 cells (passages 40–60) were cultured in Dulbecco's Modified Eagles medium (DMEM) (Gibco Invitrogen, Paisley, UK) supplemented with 20% fetal bovine serum (FBS), 1% non-essential amino acids, 1% penicillin (1000 U/mL), 1% streptomycin (1000 μg/mL) and 1% amphotericin (250 U/mL), at 37 °C under a humidified atmosphere with 5% CO_2. The cells were passaged enzymatically with 0.25% trypsin-1 mM EDTA and sub-cultured on 25 cm^2 plastic flasks at a density of 5×10^5 cells/cm^2. Culture medium was replaced every 3 days. Experiments were performed in undifferentiated cells (24 h post-seeding to prevent cell differentiation).

Human breast adenocarcinoma MCF-7 cells were kindly provided by Carlos J. Ciudad and Dr. Verónica Noé (Departamento de Bioquímica y Fisiología, Facultad de Farmacia, Universidad de Barcelona, Spain). MCF-7 cells were maintained in the same conditions as described for Caco-2 cell line.

For comparison purposes, some experiments were performed with human dermal fibroblasts that were kindly provided by Dr. Julio Montoya (Departamento de Bioquímica y Biología Molecular, Facultad de Veterinaria, Universidad de Zaragoza, 50013 Zaragoza, Spain).

2.5. Cell Viability Assay

24 h after seeding in 96-well plates at a density of 4×10^3 cells/well, cells were treated for 24 and 72 h with carbon nanomaterial dispersions (including DOX and CPT anticancer drugs in some studies), and then 3-[4,5-dimethylthiazol-2-yl]-2,5-diphenyltetrazolium bromide (MTT) assay was performed for assessing cell metabolic activity. In short, 10 µL of MTT (5 mg·mL^{-1}) were added to each 100 µL sample well and incubated for 2 h. Mitochondrial dehydrogenases of viable cells reduce the yellowish water-soluble MTT to water-insoluble formazan crystals, which are later resolubilized by replacement of the medium with DMSO, obtaining a purple colored solution. Absorbance at 540/620 nm was measured using a SPECTROstar Nano microplate reader (BMG Labtech, Ortenberg, Germany). Control values (sample wells without treatment) were set at 100% viable and all values were expressed as a percentage of the control. All experiments were performed in triplicate. In each of the three independent experiments, each sample result corresponds to 16 wells, which sums up 48 wells per sample.

2.6. Reactive Oxygen Species (ROS) Assay

The reactive oxygen species (ROS) production was assayed by the 2′,7′-dichlorofluorescein diacetate (H$_2$DCFDA) molecular probe [57,58]. The cell-permeable H$_2$DCFDA diffuses into cells and is deacetylated by cellular esterases to form 2′,7′-dichlorodihydrofluorescein (H$_2$DCF). In the presence of ROS, H$_2$DCF is rapidly oxidized to 2′,7′-dichlorofluorescein (DCF), which is highly fluorescent. Caco-2 and MCF-7 cells were seeded in 96-well plates at a density of 4×10^3 cells/well, incubated 24 h under standard cell culture conditions and then treated with nanomaterial dispersions (3 µg·mL^{-1}) for 24 h. Subsequently, cells were washed twice with PBS and incubated for 20 min with 100 µL of 20 µM H$_2$DCFDA at 37 °C for in the dark. Fluorescence intensity (ex = 485/em = 535 nm) was measured with FLUOstar Omega microplate reader (BMG Labtech). % ROS production was compared to a negative control (untreated cells) and was normalized with MTT assays at 24 h incubation. All experiments were performed in triplicate. In each of the three independent experiments, each sample result corresponds to 16 wells, which sums up 48 wells per sample.

2.7. Cell Death Study

Caco-2 and MCF-7 cells were plated in 75 cm^2 flasks at a density of 5×10^5 cells per flask and incubated 24 h under standard cell culture conditions. They were then exposed to dispersions of the tested carbon nanomaterials (3 µg·mL^{-1}) for 72 h. Each sample result corresponds to a pool of two 75 cm^2 flasks. Quantitative flow cytometry (FCT) analysis was performed using propidium iodide (PI) intake and FITC annexin V staining according to manufacturer's instruction. Briefly, cells were washed twice with phosphate saline buffer (PBS) and 100 µL of annexing V-binding buffer (10 mM HEPES/NaOH pH 7.4, 140 mM NaCl, 2.5 mM CaCl$_2$) were transferred to a 5 mL culture tube. Additions of 5 µL FITC annexin and 5 µL PI were made to each tube and then incubated for 15 min in the absence of light at room temperature. Cells were then resuspended in 400 µL of annexin V-binding buffer and analyzed with BD FACSAria flow cytometer (BD FACSDIVA version 7.0 software, accessed on 1 June 2020). Untreated cells were used as negative control and the positive control corresponds to cells treated with CPT (0.8 µg·mL^{-1}). Preliminary gating was used in flow cytometry analysis to identify the cells of interest based on the relative size and complexity of the cells, while removing debris and other events that are not of interest.

2.8. Cell Cycle Assay

Caco-2 and MCF-7 cells were plated in 75 cm^2 flasks at a density of 5×10^5 cells per flask and incubated 24 h under standard cell culture conditions. Each sample result corresponds to a pool of two 75 cm^2 flasks. They were exposed to carbon nanomaterial dispersions (3 µg·mL^{-1}) for 72 h and then washed with PBS, collected and fixed for 30 min at 4 °C and incubated with 70–80% ice-cold ethanol at −20 °C for 24 h. After washing with PBS and 5 min centrifugation at 2500× *g* rpm, cells were resuspended in PI/RNase staining buffer. PI-stained cells were analyzed for DNA content with a BD FACSArray bioanalyzer. PI fluorescence was measured in the orange range of the spectrum using a 562–588 nm band pass filter, and cell distribution was displayed on a linear scale. The percentage of cells on each cell cycle phase was determined by means of BD ModFit LT version 3.3 software (accessed on 1 June 2020).

2.9. Statistical Analysis

The experimental data were analyzed by one-way analysis of variance (ANOVA) followed by Bonferroni post-test using GraphPad Prism software (version 5.02, GraphPad Software, Inc., San Diego, CA, USA, accessed on 1 June 2020). Interval plots display 95% confidence intervals for the mean. Data were presented as means ± S.D. and differences were considered significant at $p < 0.05$.

3. Results

3.1. Characterization of Carbon Nanomaterials

Frequently, the most likely source of the apparent lack of uniformity in the results reported in the literature for in vitro and in vivo studies is the different structural and physicochemical properties of the diverse nanomaterials used. Thus, there are huge dissimilarities (i.e., length, diameter, surface defects, oxygen content, presence of impurities, etc.) among the batches employed by researchers. Therefore, thorough characterization studies of the carbon nanomaterials are required and must be taken into consideration to obtain meaningful results.

3.1.1. Transmission Electron Microscopy (TEM)

The characterization of the structural features and textural properties of the carbon nanomaterials tested here provides useful information on their interaction with drugs and cells. CNH are conical-shaped single-walled tubules that arrange into 100 nm dahlia-like assemblies (Figure 1a). The CNT used here are relatively short MWCNT (up to 1 micron in length) and ~10 nm in diameter, comprising around six concentric nanotubes (Figure 1b). TEM micrographs of two-dimensional, graphene derivatives GO and RGO (Figure 1d,e, respectively) reveal that most flakes are up to 1 micron in length as well as their high exfoliation degree. On the contrary, CNP consist of aggregates of smaller, less exfoliated graphene sheets (Figure 1c). Finally, Figure 1f shows aggregates comprising ND of about 5 nm in diameter.

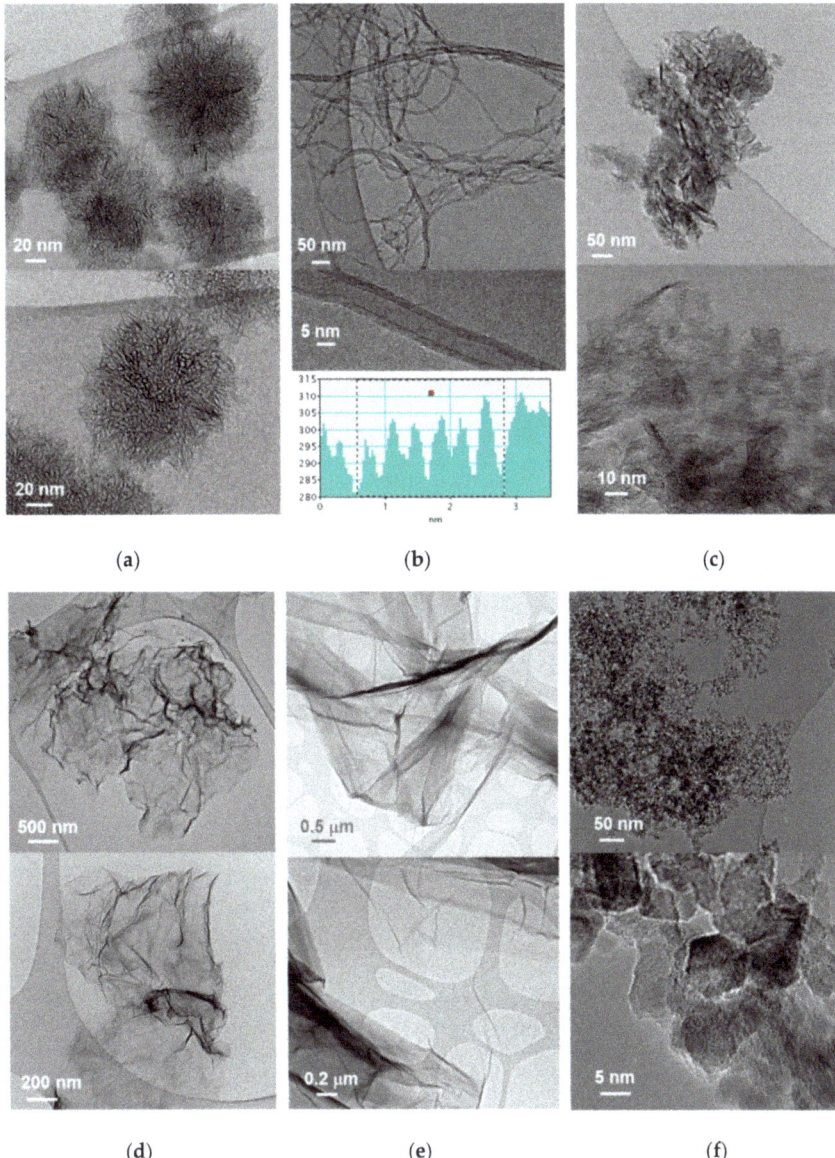

Figure 1. TEM micrographs of the tested carbon nanomaterials: (**a**) CNH, (**b**) CNT, (**c**) CNP, (**d**) RGO, (**e**) GO and (**f**) ND. Inset in (**b**) shows image analysis of a MWCNT with six concentric nanotubes.

3.1.2. Photoelectron Spectroscopy (XPS)

X-ray photoelectron spectroscopy (XPS) provides important hints of the surface chemistry of the tested carbon nanomaterials (Figure S1). The ratio of oxygen and carbon atoms was calculated from the O1s and C1s peaks, and the results of the quantitative surface analysis are summarized in Table 1. XPS spectra of CNH, CNT, CNP and RGO are quite similar and correspond to C sp^2-based nanomaterials materials, with a low O:C ratio and, therefore, are highly hydrophobic. In contrast, GO has a significant high oxygen content (49.2 at.%), as it contains abundant oxygen-containing functional groups, which provide enhanced hydrophilicity. Although O content in ND is not as high as in GO, ND are known to disperse easily in polar solvents, as it will be commented later in the Discussion section. No significant transition metal contamination was observed in XPS spectra.

Table 1. Surface chemical analysis (at.%) of the carbon nanomaterials, obtained from XPS spectra.

At.%	CNH	CNT	CNP	RGO	GO	ND [1]
C	92.5	94.7	89.7	82.0	50.8	80.5
O	7.5	5.3	10.3	18.0	49.2	16.5

[1] For ND, at.% N is 3.0, calculated from the N1s peak in XPS spectra.

3.1.3. Specific Surface Area

Specific surface area for carbon nanomaterials in powder, determined using N_2 adsorption and BET method, are shown in Table 2.

Table 2. Specific surface area of carbon nanomaterials determined by BET method.

	CNH	CNT	CNP	RGO	GO	ND
Specific surface area ($m^2 \cdot g^{-1}$)	438.6	305.5	701.8	344.6	103.3	331.4

The largest specific surface area value corresponds to CNP. While these values correspond to powder samples, sonication-assisted dispersion in solution significantly increases surface area, which is particularly relevant when it comes to GO exfoliation.

3.2. Dispersion of Carbon Nanomaterials

For efficient cellular uptake of carbon nanomaterials, it is necessary that that they remain dispersed and not aggregate in culture medium. Non-ionic polyether surfactants, such as poloxamer triblock copolymers (known also by the trade name Pluronic®), are frequently used as dispersants to prepare various nanoparticle suspensions, especially with hydrophobic nanoparticles, such as CNT and related materials. Pluronics are amphiphilic molecules that comprise two polyethylene glycol (PEG) blocks and one polypropylene glycol (PPG) block of various sizes and are frequently used for in vitro and in vivo nanotoxicity studies because they are considered non-toxic dispersants. Thus, the US Food and Drug Administration (FDA) has approved various Pluronic polymers for pharmaceutical usage and even intravenous administration [59,60]. However, it is known that Pluronics can be degraded during sonication, depending on sonication time, power, and frequency conditions, as the collapse of cavitation bubbles generated during sonication can create sufficient heat, pressures, and shear forces to degrade polymers containing PEG, PPG or both. It is therefore important to assess whether sonication of dispersants themselves contribute to the toxicity of sonicated nanomaterial suspensions so as not to misinterpret toxicity results [61]. Figure S2 shows that F-127 decreased MCF-7 and Caco-2 cell viability at high concentration. Thus, F-127 at low concentration (15 $\mu g \cdot mL^{-1}$) and short bath sonication time (<10 min) was used here to assist in carbon nanomaterials dispersion in cell culture medium, while avoiding the generation of toxic degradation products. Moreover, it is well documented in the literature that above critical micelle concentration (CMC), Pluronics form nano-sized micellar structures which can act as drug nanocarriers, showing higher anticancer activity as compared to free drug [59,62]. It was also checked here that neither DOX nor CPT anticancer activity was enhanced due to drug encapsulation in F-127 micellar structures at the low F-127 concentration used here (Figure S3). Therefore, any improvement achieved in this study in cell killing ability over free drug against cancer cells can be attributed to the drug-nanocarrier complex.

3.3. Carbon Nanomaterials Toxicity Assessment

Cell viability assay, apoptosis detection, cell cycle analysis and ROS production assay are useful in vitro methods for the assessment of toxicity of nanomaterials.

3.3.1. Cell Viability Assay

Figure 2 shows cell viability assays on carbon nanomaterial treatment at 3 µg·mL^{-1} after 24 and 72 h for Caco-2 and MCF-7 cell lines. Also results at 0.6 µg·mL^{-1} can be found in Figure S4. The MTT assays showed dose-dependence on both Caco-2 and MCF-7 cell lines. Cell viability followed the order of CNP < CNH < RGO < CNT < GO < ND. The decrease in cell viability was more pronounced for the Caco-2 cell line. No significant cell viability decrease was observed in Figure 2 for GO and ND at 3 µg·mL^{-1} (and also for CNT, at 0.6 µg·mL^{-1}, as shown in Figure S4).

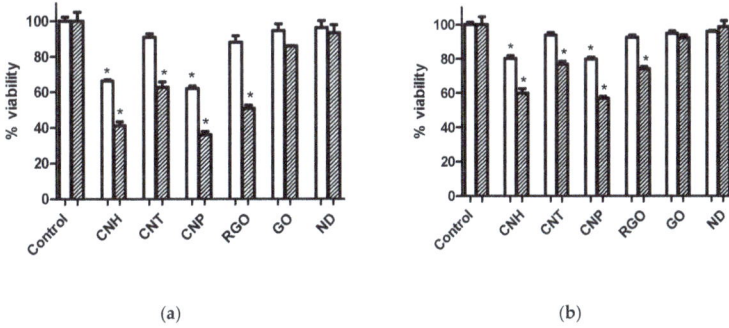

Figure 2. Cell viability assays after 24 h (white) and 72 h (striped) of incubation with various carbon nanomaterials at 3 µg·mL^{-1} showing differential effects on (**a**) Caco-2 and (**b**) MCF-7 cells. Values that are significantly different from the control ($p < 0.05$) are denoted with asterisk (*). Untreated cells were used as control.

Interestingly, MTT studies on human dermal fibroblast cells (Figure S5), as example of healthy cells, provided the same viability sequence after carbon nanomaterial treatment, showing less effects than on Caco-2 cells and very similar to those on MCF-7 cells.

3.3.2. Reactive Oxygen Species (ROS) Assays

According to Figure 3, CNP produced the highest levels of ROS, more noticeably for Caco-2 cells.

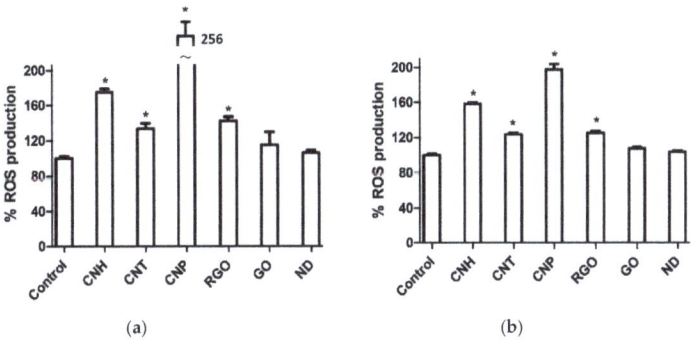

Figure 3. ROS generation on (**a**) Caco-2 and (**b**) MCF-7 cells upon incubation with nanomaterial dispersions (3 µg·mL^{-1}) for 24 h. Significant results as compared to untreated control cells are marked by asterisk * for p-value < 0.05.

3.3.3. Cell Death Study

Figure 4 shows quantitative flow cytometry analyses for the Caco-2 and MCF-7 cell lines, treated 72 h with CNH at 3 µg·mL^{-1}, which showed the highest values of late apoptosis/necrosis among the carbon nanomaterials tested here. Experiments were performed as described in Section 2.7, and interpreted as follows: the percentage of viable cells is shown in the lower left quadrant (annexin V$^-$/PI$^-$), of early apoptotic cells in the lower right quadrant (annexin V$^+$/PI$^-$), and of late apoptotic and necrotic cells in the upper right quadrant (annexin V$^+$/PI$^+$). Additional results for CNT, CNP and RGO are compared in Figure S6.

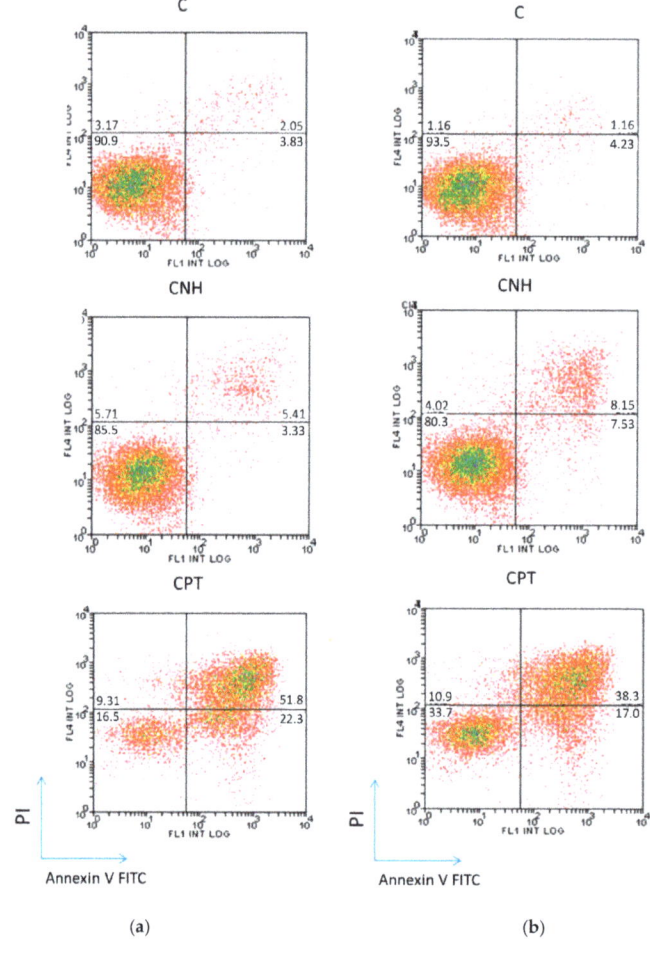

Figure 4. Annexin V/Propidium iodide assay, providing quantitative information about living (lower left quadrant), early apoptotic (lower right quadrant) and late apoptotic and necrotic (upper right quadrant) cells for (**a**) Caco-2 and (**b**) MCF-7 cell lines, treated with CNH at 3 µg·mL^{-1} for 72 h. Data are presented as percentage of the cell population. Untreated cells, denoted as C, were used as negative control, and CPT (0.8 µg·mL^{-1}) treated cells were used as positive control.

3.3.4. Cell Cycle Analysis

Figure 5 shows flow cytometric analysis of Caco-2 and MCF-7 cell cycle after treatment with CNH at 3 µg·mL^{-1} for 72 h. Additional results for CNT, CNP and RGO are compared in Figure S7.

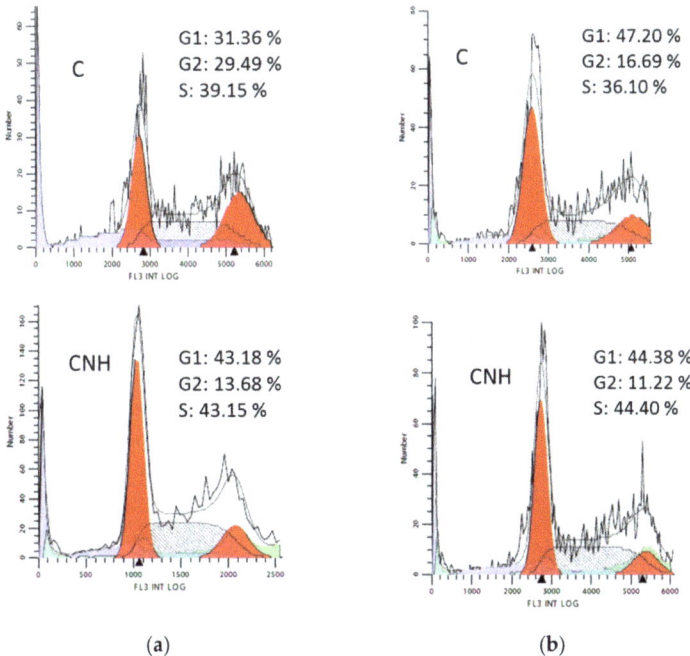

(a) (b)

Figure 5. Flow cytometric analysis of (**a**) Caco-2 and (**b**) MCF-7 cell cycle after treatment with CNH at 3 µg·mL^{-1} for 72 h. C denotes untreated control cells.

3.4. Carbon Nanomaterials as Anticancer Drug Nanocarriers

The potential in the field of drug delivery as nanocarriers of anticancer drugs of the four carbon nanomaterials studied here that showed the lowest effect on the cells (ND, GO, CNT and RGO) was compared in Figure 6. MTT assays were performed on Caco-2 cells at two drug concentrations, 0.2 and 0.8 µg·mL^{-1}. Carbon nanomaterial concentration was chosen as low as 0.6 µg·mL^{-1}, so that the observed decrease in cell viability could be attributable to the improved DOX or CPT efficacy when loaded on carbon nanomaterial nanocarriers rather than to any inherent toxicity of carbon nanomaterials. CPT showed more potent cytotoxic activity than DOX against both cancer cells (Figure 6). Carbon nanomaterial/drug complexes resulted in improved anticancer activity compared to that of the free drug. For CPT, the improvement follows the sequence ND < GO < CNT < RGO. For DOX, the sequence is the opposite (Figure 6). Thus, CNT and RGO showed significant enhanced anticancer activity compared to the free drug, but ND showed a significant improvement when it comes to DOX.

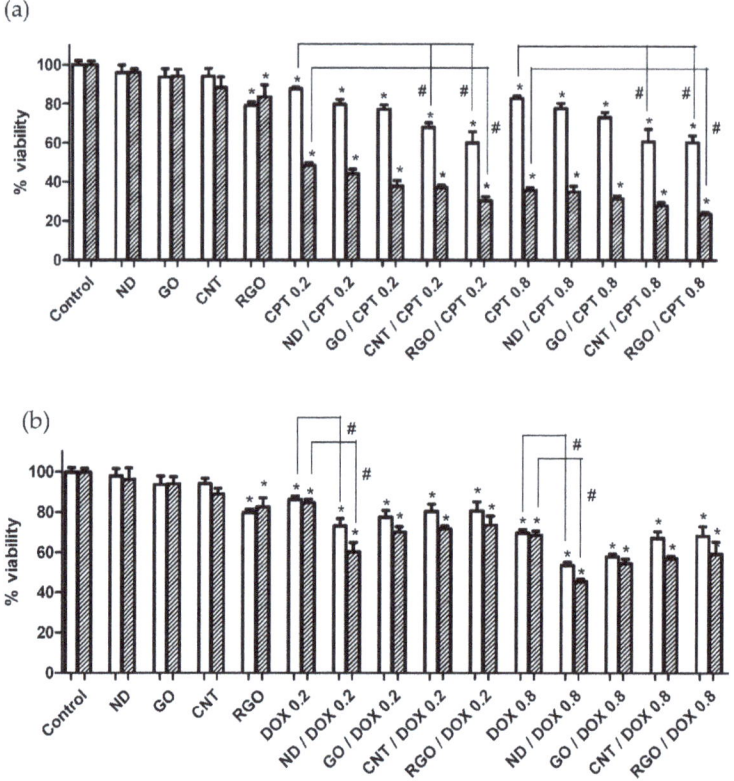

Figure 6. Cell viability assays after 24 h (white) and 72 h (striped) of incubation of Caco-2 cells with ND, GO, CNT and RGO at 0.6 µg·mL^{-1} concentration, free drug CPT (**a**) and DOX (**b**) at both 0.2 and 0.8 µg·mL^{-1} concentrations, and CPT- (**a**) and DOX- (**b**) loaded carbon nanomaterials. (* and # represent significance at $p < 0.05$ when compared to untreated control cells and free drug-treated cells, respectively).

4. Discussion

CNT are a type of hollow one-dimensional (1D) carbon-based nanomaterial consisting of a graphene sheet rolled up to form a cylindrical structure with sp^2 hybridized carbon atoms. CNT are classified into single-walled carbon nanotubes (SWCNT) and multi-walled carbon nanotubes (MWCNT), have high aspect ratios and needle-like shapes [63]. Comparing the two types, there has been a major debate over whether SWCNTs or MWCNTs generate more toxicity. Some research groups have reported that SWCNT cause more apoptosis than MWCNT, as they are more agglomerated [64–66]. Moreover, short CNT were found to be less toxic than longer CNT, which is comparable with the observed toxicity of asbestos [44,65,67,68]. CNT used here are MWCNT, with relatively short length (mean length 1 µm, Figure 1b), so that low toxicity is expected. The purity of this CNT material is high (>95.0%) so no significant toxicity should result from any traces of the transition metal nanoparticles used during CNT production.

Single-walled carbon nanohorns (SWCNH) are horn-shaped single-walled tubules with cone angles of approximately 20° that usually form aggregates with diameters of 80–100 nm [69,70] with a "dahlia-like" shape, as shown in Figure 1a. They are produced essentially metal-free and with high purity [71]. Their use in biomedical applications is still at a preliminary stage. SWCNH used here were produced without any catalyst by direct vaporization of graphite, as described in Section 2.1.

Another category of carbon nanomaterial is graphene, a two-dimensional (2D) sp^2-bonded carbon sheet in a honeycomb structure and therefore, pristine graphene is hydrophobic in nature. On the contrary, GO contains abundant epoxy and hydroxyl functional groups attached to the basal plane and carboxylic groups attached to the edges, that disrupt the π conjugation, providing enhanced hydrophilicity which even enables the efficient dispersion in aqueous media. π conjugation and therefore hydrophobicity are partially restored upon reduction in RGO [72,73]. Size and morphological characteristics of graphene derivatives studied here, CNP, RGO and GO, are shown in Figure 1c–e.

As another important member of carbon nanomaterial family, ND consist of a highly ordered diamond core covered by a layer of functional groups on the surface, such as carboxyl, lactone, hydroxy and ketone, which stabilizes the particle by terminating the dangling bonds [74,75]. ND produced by detonation method are extremely tiny particles with average diameter between 4–6 nm (Figure 1f). ND are becoming increasingly useful in therapeutic and diagnostic applications due to their biocompatibility, scalability, and easy surface modification [76,77].

According to the XPS results summarized in Table 1, CNH, CNT, CNP and RGO nanomaterials have a low O:C ratio and can be considered as hydrophobic and difficult to disperse in polar solvents. On the contrary, GO has a remarkable high oxygen content and can be considered hydrophilic. Although oxygen content in ND is lower than in GO (Table 1), ND are known to disperse easily in polar solvents, which is due to the hydrophilic functional groups on the outer shell. No significant transition metal contamination for the tested nanomaterials was observed by TEM and XPS. Thus, we can claim that the toxicological effects of metal impurities in these nanomaterials are negligible.

Amphiphilic F-127 was used here to assist the dispersion of carbon nanomaterials in cell culture media through noncovalent functionalization, which involves the coating of the carbon nanomaterials with hydrophobic PPG motifs anchored onto the material surface, with the hydrophilic PEG ends extending to the aqueous solution and enabling the stability of the material in aqueous media.

Results of MTT assays upon treatment with carbon nanomaterials at 3 µg·mL^{-1} (Figure 2) and 0.6 µg·mL^{-1} (Figure S4) show that the cell viability was cellular type, time and dose-dependent. Viability decrease was more pronounced on the highly active metabolically Caco-2 cells. Cell viability follows the order CNP < CNH < RGO < CNT < GO < ND for both Caco-2 and MCF-7 line cells. The sequence in cell viability that resulted from the MTT assays for the different carbon nanomaterials tested here can be explained taking into account the surface chemistry of carbon nanomaterials. Thus, oxygen functional groups on the surface of carbon nanomaterials shield the hydrophobic domains. Two groups of carbon nanomaterials can be distinguished here, the hydrophilic ones, ND and GO, which present low effect on cells, and the hydrophobic ones, CNT, RGO, CNH and CNP, which inhibited cell viability in more extent. The highest viability values correspond to ND, whose surface is rich in functional groups, which make them ideal nanocarriers for building drug delivery systems. However, as it will later be discussed, ND efficiently load hydrophilic drugs, such as DOX, which readily attach to their functional groups on their surface, rather than hydrophobic drugs, such as CPT.

Figure 3 shows that, compared to the other carbon nanomaterials, CNP produced the highest ROS levels, more pronounced for Caco-2 cells. We also found enhancement of ROS levels for cells treated with CNH respect to those treated with CNT and RGO. No significant ROS level alterations were however observed for ND and GO.

As for the apoptosis study, the combination of annexin V and PI has been used to discriminate early apoptotic cells from late apoptotic and necrotic ones. Results collected in Figure 4 and Figure S6 show that the hydrophobic carbon nanomaterials induced late apoptosis/necrosis for both Caco-2 and MCF-7 cells, being more pronounced for CNH. Anticancer drug CPT at 0.8 µg·mL^{-1}, whose toxicity was much larger than that of all carbon nanomaterials studied here, was used as positive control.

The effect of carbon nanomaterials on cell cycle progression in Caco-2 and MCF-7 cells is shown in Figure 5 and Figure S7. Cytometric analysis showed no significant differences in the percentage of cells in the individual phases of the cell cycle for all the tested carbon nanomaterials and untreated cells, particularly for MCF-7 cells

Taking these results all together, we conclude that ND and GO show low toxicity, which is due to the oxygenated functional groups on their surface that shield the hydrophobic domains. On the other hand, CNH and CNP induce Caco-2 and MCF-7 late apoptosis/necrosis and enhanced ROS levels, which could be associated with the higher decrease in cell viability, compared to other hydrophobic carbon nanomaterials such as CNT and RGO. This could probably be due to the "dahlia-like" CNH morphology, consisting of small structures containing sharp conical ends, that may produce damage to cells, as well as the sharp edges of the highly fragmented CNP platelets [78,79]. Finally, CNP were found to induce the most elevated levels of ROS, which would contribute to the highest observed decrease in cell viability. The effect is noticeably more pronounced on the more rapidly dividing Caco-2 cells.

It is worth noting that CNH was reported to inhibit proliferation of human liver cell lines and promoted apoptosis [80]. In contrast, other authors reported low toxicity for CNH [69,81,82]. It has to be noted that low toxicity reports correspond to CNH synthesis methods leading to oxidized CNH, such as CO_2 laser ablation or arc discharge. Thus, oxygen functional groups on the surface would shield hydrophobic carbon domains from interactions with cellular membranes. CNH studied here were produced by direct vaporization of graphite, with very low O content, which was confirmed by XPS (Table 1). This highlights the importance of carbon nanomaterial source when drawing meaningful conclusions from toxicity studies.

Finally, the potential of carbon nanomaterials materials in the field of drug delivery of anticancer drug was compared here. Drug delivery systems based on noncovalent interactions have several advantages compared with covalent conjugation. Thus, extra steps required in chemical conjugations are not necessary. Also, because the drug structure is not chemically altered, drug molecules released from such delivery systems are expected to exert their predicted pharmacological effects. Many clinically used chemical drugs possess aromatic rings, such as DOX, playing the π–π stacking interactions the major role in drug delivery systems [83]. It is known that the loading efficiency for DOX decreases when using CNTs with higher levels of PEGylation, due to the increased hydrophilicity of the surface. Furthermore, faster release rates of DOX were observed for these higher PEGylated CNTs owing to the lower affinity of DOX to the PEGylated CNT [84].

Due to their sp^2 carbon structure and inherent hydrophobic nature of carbon nanomaterials, all of them (except ND) are capable of establishing noncovalent π–π stacking interactions for the formation of anticancer DOX and CPT complexes. As for the hydrophobic drug CPT (Figure 7a), the more hydrophobic the carbon nanomaterial is, the more C sp^2 domains has and the more efficient is the loading of CPT, through strong π–π interactions, which explains the results shown in Figure 6. Thus, the highest improvement in CPT anticancer activity compared to the free drug was observed for RGO and CNT nanocarriers.

On the other hand, because of its high surface free energy, ND rarely exist as a single particle, and usually form clusters of tens to hundreds of nanometers, even when they are dispersed in a solution by strong ultrasonication. Drug molecules can be assembled on the surface of ND clusters or in the nanoscale pores inside the ND clusters (Figure 7b) by noncovalent interactions [77,85]. The highest improvement in DOX activity compared to that of the free drug was observed for ND. However, ND were not efficient in loading the more hydrophobic drug, CPT. Results for higher ND concentration up to 20 $\mu g \cdot mL^{-1}$ shown in Figure S8 show that the efficiency was worse than those at 0.6 $\mu g \cdot mL^{-1}$, probably due to ND aggregation forming higher size clusters, which offer less surface area for drug loading and more difficulty to enter the cells.

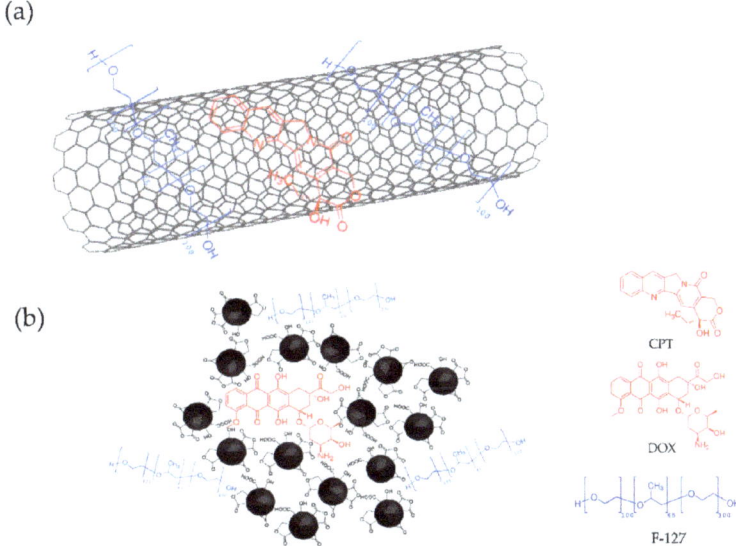

Figure 7. Schema of mechanism of interaction between drugs (in red) and carbon nanomaterials (in black): (**a**) CPT/CNT and (**b**) DOX/ND. (F-127 is depicted in blue).

Carbon nanomaterials display unique physicochemical properties making them potentially useful for bioapplications and competitive when compared to micelles, polymeric nanoparticles, dendrimers, and liposomes, to name a few. Thus, they offer high surface area for multiple drug adsorption through π–π stacking interactions and, as for ND, drugs bound to the abundant functional groups on their surface show enhanced chemotherapeutic efficacy. Much research activity has been devoted to perform in vivo experiments, either by systemic administration and localized drug delivery strategies [86]. Remarkably, carbon nanomaterials have also received much attention in imaging and diagnostics. Thus, due to their strong absorption in the IR or NIR regions, they can be used in cancer photothermal therapy (PTT). Also, they are useful in fluorescence [87,88] and photoacoustic imaging (PAI) [89–91]. Intrinsic carbon nanomaterial Raman vibrations allow monitoring their in vivo distribution and metabolism [32,33]. ND presenting nitrogen-vacancy centers have intrinsic fluorescence properties, and therefore are interesting tools for imaging and diagnostics [75]. Finally, carbon-based nanomaterials are emerging as potential candidates for the development of synthetic scaffolds in tissue engineering [92–95].

Long-term fate of carbon nanomaterials has been the subject of much concern and the origin of much skepticism surrounding their in vivo applications, as are presumed to be biopersistent. Despite discrepancies in findings on the clearance mechanism, majority of the studies have suggested that increasing the degree of functionalization enhanced renal clearance, while lower functionalization promoted RES accumulation (i.e., liver and spleen) [96]. Several groups have reported that carbon-based nanomaterials are susceptible to biodegradation as a result of the key role played by the immune system [97].

5. Conclusions

Cytotoxicity evaluation after 24 h and 72 h of incubation with various carbon nanomaterials shows differential effects on Caco-2 and MCF-7 cells. Cell viability followed the order CNP < CNH < RGO < CNT < GO < ND, being more pronounced in the more rapidly dividing Caco-2 cells. ND and GO showed the lowest toxicity, due to the presence of oxygen functional groups on carbon nanomaterial surface, that shield the hydrophobic carbon domains. High hydrophobicity, together with the morphology containing sharp conical ends in CNH and sharp edges in CNP would account for the high cell viability decrease, enhanced ROS level and apoptosis/necrosis. Remarkable high ROS levels were obtained for CNP, more pronounced on Caco-2 cells.

There is a lack of ROS generation from both cell lines after incubation with ND, as well as the lowest apoptosis values, which further supports that ND provide the lowest toxicity among the carbon nanomaterials tested here, which make them an ideal carrier for designing drug delivery systems. ND form clusters of tens to hundreds of nanometers, wherein drugs can be loaded by interaction with their surface functional groups, and therefore ND will be much more efficient in loading hydrophilic drugs, such as DOX, which readily attach to their functional groups on their surface, rather than hydrophobic drugs. In contrast, CNT and RGO, which also have low toxicity among the hydrophobic carbon nanomaterials tested here, offer available surface area for π–π interactions with aromatic rings, leading to high CPT loading efficiency, due to the strong π–π stacking interactions formed with CPT. Remarkably, CPT is a more potent anticancer agent than DOX, so developing new drug delivery systems for CPT is of high interest.

Several obstacles must be overcome before carbon nanomaterials can be suitable for clinical use. The major challenge and current limitation in this area is still the potential long-term toxicity concerns of carbon nanomaterials. Comparative in vitro studies of cytotoxicity of carbon nanomaterial synthesized from different sources are needed as screening and risk-to-benefit assessment, together with drug loading efficiency studies, to further develop advanced multi-functional carbon nanomaterials for cancer theranostic applications.

Supplementary Materials: The following are available online at http://www.mdpi.com/2079-4991/10/8/1617/s1, Figure S1: High resolution XPS spectra of carbon nanomaterials; Figure S2: Cell viability assays showing the effect of F-127 at two concentrations; Figure S3: Cell viability assays after treatment with DOX and CPT, in the presence and absence of F-127; Figure S4: Cell viability assays after treatment with carbon nanomaterials at 0.6 µg·mL^{-1}; Figure S5: Cell viability assays after treatment with carbon nanomaterials at 3.0 µg·mL^{-1} on human dermal fibroblast cells; Figure S6: Cell death study comparing different hydrophobic carbon nanomaterials; Figure S7: Cell cycle analysis comparing different hydrophobic carbon nanomaterials; Figure S8: Cell viability assays after treatment with ND at several concentrations, DOX and CPT.

Author Contributions: R.G. and M.J.R.-Y. were responsible for the overall direction of the research. V.L.C. and E.M. contributed with nanomaterial characterization supervised by R.G., M.P. and T.H.-C. performed cell culture experiments supervised by R.G. and M.J.R.-Y., R.G., M.J.R.-Y. and J.O. analyzed the obtained data. R.G. drafted the manuscript. All authors have read and agreed to the published version of the manuscript.

Funding: This research was supported by the Diputación General de Aragón (project E25_20R); the Spanish Ministry of Economy and Innovation under Grant (SAF 2016-75441-R); Aragón Regional Government (B16-17R, Fondos FEDER "Otra manera de hacer Europa"); CIBEROBN under Grant (CB06/03/1012) of the Instituto de Salud Carlos III and European Grant Interreg/SUDOE (Redvalue, SOE1/PI/E0123).

Acknowledgments: The TEM microscopy work has been conducted in the "Laboratorio de Microscopías Avanzadas" at "Instituto de Nanociencia de Aragón—Universidad de Zaragoza". The authors acknowledge the LMA-INA for offering access to their instruments and expertise. Quantitative flow cytometric studies were performed at "Centro de Investigación Biomédica de Aragón", Spain (http://www.iacs.aragon.es). The authors thank Alfonso Ibarra, Rodrigo Fernández-Pacheco, Ariel Ramírez-Labrada, Elvira Aylón and Isaías Fernández for valuable technical support and fruitful discussions.

Conflicts of Interest: The authors declare no conflict of interest.

References

1. Wong, B.S.; Yoong, S.L.; Jagusiak, A.; Panczyk, T.; Ho, H.K.; Ang, W.H.; Pastorin, G. Carbon nanotubes for delivery of small molecule drugs. *Adv. Drug Deliv. Rev.* **2013**, *65*, 1964–2015. [CrossRef] [PubMed]
2. Liu, Z.; Robinson, J.T.; Tabakman, S.M.; Yang, K.; Dai, H. Carbon materials for drug delivery & cancer therapy. *Mater. Today* **2011**, *14*, 316–323. [CrossRef]
3. Chen, D.; Dougherty, C.A.; Zhu, K.; Hong, H. Theranostic applications of carbon nanomaterials in cancer: Focus on imaging and cargo delivery. *J. Control. Release* **2015**, *210*, 230–245. [CrossRef] [PubMed]
4. Kabanov, A.V. Polymer genomics: An insight into pharmacology and toxicology of nanomedicines. *Adv. Drug Deliv. Rev.* **2006**, *58*, 1597–1621. [CrossRef] [PubMed]

5. Lanone, S.; Boczkowski, J. Biomedical applications and potential health risks of nanomaterials: Molecular mechanisms. *Curr. Mol. Med.* **2006**, *6*, 651–663. [CrossRef] [PubMed]
6. Maeda, H. The enhanced permeability and retention (EPR) effect in tumor vasculature: The key role of tumor-selective macromolecular drug targeting. *Adv. Enzyme Regul.* **2001**, *41*, 189–207. [CrossRef]
7. Eatemadi, A.; Daraee, H.; Karimkhanloo, H.; Kouhi, M.; Zarghami, N.; Akbarzadeh, A.; Abasi, M.; Hanifehpour, Y.; Joo, S.W. Carbon nanotubes: Properties, synthesis, purification, and medical applications. *Nanoscale Res. Lett.* **2014**, *9*, 393–405. [CrossRef]
8. Clancy, A.J.; White, E.R.; Tay, H.H.; Yau, H.C.; Shaffer, M.S.P. Systematic comparison of conventional and reductive single-walled carbon nanotube purifications. *Carbon* **2016**, *108*, 423–432. [CrossRef]
9. Dumortier, H.; Lacotte, S.; Pastorin, G.; Marega, R.; Wu, W.; Bonifazi, D.; Briand, J.P.; Prato, M.; Muller, S.; Bianco, A. Functionalized carbon nanotubes are non-cytotoxic and preserve the functionality of primary immune cells. *Nano Lett.* **2006**, *6*, 1522–1528. [CrossRef]
10. Mishra, P.; Mishra, B.; Dey, R.K. PEGylation in anti-cancer therapy: An overview. *Asian J. Pharm. Sci.* **2016**, *11*, 337–348. [CrossRef]
11. Greenwald, R.B.; Choe, Y.H.; McGuire, J.; Conover, C.D. Effective drug delivery by PEGylated drug conjugates. *Adv. Drug Deliv. Rev.* **2003**, *55*, 217–250. [CrossRef]
12. Kolate, A.; Baradia, D.; Patil, S.; Vhora, I.; Kore, G.; Misra, A. PEG—A versatile conjugating ligand for drugs and drug delivery systems. *J. Control. Release* **2014**, *192*, 67–81. [CrossRef] [PubMed]
13. Miao, W.; Shim, G.; Lee, S.; Lee, S.; Choe, Y.S.; Oh, Y.K. Safety and tumor tissue accumulation of PEGylated graphene oxide nanosheets for co-delivery of anticancer drug and photosensitizer. *Biomaterials* **2013**, *34*, 3402–3410. [CrossRef] [PubMed]
14. Zhang, X.; Meng, L.; Lu, Q.; Fei, Z.; Dyson, P.J. Targeted delivery and controlled release of doxorubicin to cancer cells using modified single wall carbon nanotubes. *Biomaterials* **2009**, *30*, 6041–6047. [CrossRef] [PubMed]
15. Zhang, Q.; Wu, Z.; Li, N.; Pu, Y.; Wang, B.; Zhang, T.; Tao, J. Advanced review of graphene-based nanomaterials in drug delivery systems: Synthesis, modification, toxicity and application. *Mater. Sci. Eng. C* **2017**, *77*, 1363–1375. [CrossRef]
16. Allen, T.M. Ligand-targeted therapeutics in anticancer therapy. *Nat. Rev. Cancer* **2002**, *2*, 750–763. [CrossRef]
17. Schrama, D.; Reisfeld, R.A.; Becker, J.C. Antibody targeted drugs as cancer therapeutics. *Nat. Rev. Drug Discov.* **2006**, *5*, 147–159. [CrossRef]
18. Heister, E.; Neves, V.; Tîlmaciu, C.; Lipert, K.; Sanz-Beltrán, V.; Coley, H.M.; Silva, S.R.P.; McFadden, J. Triple functionalisation of single-walled carbon nanotubes with doxorubicin, a monoclonal antibody, and a fluorescent marker for targeted cancer therapy. *Carbon* **2009**, *47*, 2152–2160. [CrossRef]
19. Yang, D.; Feng, L.; Dougherty, C.A.; Luker, K.E.; Chen, D.; Cauble, M.A.; Banaszak Holl, M.M.; Luker, G.D.; Ross, B.D.; Liu, Z.; et al. In vivo targeting of metastatic breast cancer via tumor vasculature-specific nano-graphene oxide. *Biomaterials* **2016**, *104*, 361–371. [CrossRef]
20. Li, R.; Wu, R.; Zhao, L.; Wu, M.; Yang, L.; Zou, H. P-glycoprotein antibody functionalized carbon nanotube overcomes the multidrug resistance of human leukemia cells. *ACS Nano* **2010**, *4*, 1399–1408. [CrossRef]
21. Mura, S.; Nicolas, J.; Couvreur, P. Stimuli-responsive nanocarriers for drug delivery. *Nat. Mater.* **2013**, *12*, 991–1003. [CrossRef]
22. Janib, S.M.; Moses, A.S.; MacKay, J.A. Imaging and drug delivery using theranostic nanoparticles. *Adv. Drug Deliv. Rev.* **2010**, *62*, 1052–1063. [CrossRef]
23. Koo, H.; Huh, M.S.; Sun, I.C.; Yuk, S.H.; Choi, K.; Kim, K.; Kwon, C. In vivo targeted delivery of nanoparticles for theranosis. *Acc. Chem. Res.* **2011**, *44*, 1018–1028. [CrossRef]
24. Huang, P.; Xu, C.; Lin, J.; Wang, C.; Wang, X.; Zhang, C.; Zhou, X.; Guo, S.; Cui, D. Folic Acid-conjugated graphene oxide loaded with photosensitizers for targeting photodynamic therapy. *Theranostics* **2011**, *1*, 240–250. [CrossRef]
25. Mendes, R.G.; Bachmatiuk, A.; Büchner, B.; Cuniberti, G.; Rümmeli, M.H. Carbon nanostructures as multi-functional drug delivery platforms. *J. Mater. Chem. B* **2013**, *1*, 401–428. [CrossRef]
26. Prato, M.; Kostarelos, K.; Bianco, A. Functionalized carbon nanotubes in drug design and discovery. *Acc. Chem. Res.* **2008**, *41*, 60–68. [CrossRef]

27. Singh, R.; Pantarotto, D.; Lacerda, L.; Pastorin, G.; Klumpp, C.; Prato, M.; Bianco, A.; Kostarelos, K. Tissue biodistribution and blood clearance rates of intravenously administered carbon nanotube radiotracers. *Proc. Natl. Acad. Sci. USA* **2006**, *103*, 3357–3362. [CrossRef]
28. Wu, W.; Li, R.; Bian, X.; Zhu, Z.; Ding, D.; Li, X.; Jia, Z.; Jiang, X.; Hu, Y. Covalently combining carbon nanotubes with anticancer agent: Preparation and antitumor activity. *ACS Nano* **2009**, *3*, 2740–2750. [CrossRef]
29. Jasim, D.A.; Ménard-Moyon, C.; Bégin, D.; Bianco, A.; Kostarelos, K. Tissue distribution and urinary excretion of intravenously administered chemically functionalized graphene oxide sheets. *Chem. Sci.* **2015**, *6*, 3952–3964. [CrossRef]
30. Das, M.; Datir, S.R.; Singh, R.P.; Jain, S. Augmented anticancer activity of a targeted, intracellularly activatable, theranostic nanomedicine based on fluorescent and radiolabeled, methotrexate-folic acid-multiwalled carbon nanotube conjugate. *Mol. Pharm.* **2013**, *10*, 2543–2557. [CrossRef]
31. Yang, K.; Zhang, S.; Zhang, G.; Sun, X.; Lee, S.T.; Liu, Z. Graphene in mice: Ultrahigh in vivo tumor uptake and efficient photothermal therapy. *Nano Lett.* **2010**, *10*, 3318–3323. [CrossRef] [PubMed]
32. Liu, A.; Davis, C.; Cai, W.; He, L.; Chen, X.; Dai, H. Circulation and long-term fate of functionalized, biocompatible single-walled carbon nanotubes in mice probed by Raman spectroscopy. *Proc. Natl. Acad. Sci. USA* **2008**, *105*, 1410–1415. [CrossRef]
33. Liu, Z.; Li, X.; Tabakman, S.M.; Jiang, K.; Fan, S.; Dai, H. Multiplexed multicolor Raman imaging of live cells with isotopically modified single walled carbon nanotubes. *J. Am. Chem. Soc.* **2008**, *130*, 13540–13541. [CrossRef]
34. Moon, H.K.; Lee, S.H.; Choi, H.C. In vivo near-infrared mediated tumor destruction by photothermal effect of carbon nanotubes. *ACS Nano* **2009**, *3*, 3707–3713. [CrossRef] [PubMed]
35. Yang, K.; Feng, L.; Shi, X.; Liu, Z. Nano-graphene in biomedicine: Theranostic applications. *Chem. Soc. Rev.* **2013**, *42*, 530–547. [CrossRef] [PubMed]
36. Chen, J.; Liu, H.; Zhao, C.; Qin, G.; Xi, G.; Li, T.; Wang, X.; Chen, T. One-step reduction and PEGylation of graphene oxide for photothermally controlled drug delivery. *Biomaterials* **2014**, *35*, 4986–4995. [CrossRef] [PubMed]
37. Parhi, P.; Mohanty, C.; Sahoo, S.K. Nanotechnology-based combinational drug delivery: An emerging approach for cancer therapy. *Drug Discov. Today* **2012**, *17*, 1044–1052. [CrossRef]
38. Jabr-Milane, L.S.; Van Vlerken, L.E.; Yadav, S.; Amiji, M.M. Multi-functional nanocarriers to overcome tumor drug resistance. *Cancer Treat Rev.* **2008**, *34*, 592–602. [CrossRef]
39. Lacerda, L.; Herrero, M.A.; Venner, K.; Bianco, A.; Prato, M.; Kostarelos, K. Carbon-nanotube shape and individualization critical for renal excretion. *Small* **2008**, *4*, 1130–1132. [CrossRef]
40. Ruggiero, A.; Villa, C.H.; Bander, E.; Rey, D.A.; Bergkvist, M.; Batt, C.A.; Manova-Todorova, K.; Deen, W.M.; Scheinberg, D.A.; McDevitt, M.R. Paradoxical glomerular filtration of carbon nanotubes. *Proc. Natl. Acad. Sci. USA* **2010**, *107*, 12369–12374. [CrossRef]
41. Kiew, S.F.; Kiew, L.V.; Lee, H.B.; Imae, T.; Chung, L.Y. Assessing biocompatibility of graphene oxide-based nanocarriers: A review. *J. Control. Release* **2016**, *226*, 217–228. [CrossRef] [PubMed]
42. Kostarelos, K.; Bianco, A.; Prato, M. Promises, facts and challenges for carbon nanotubes in imaging and therapeutics. *Nat. Nanotechnol.* **2009**, *4*, 627–633. [CrossRef] [PubMed]
43. Jastrzębska, A.M.; Kurtycz, P.; Olszyna, A.R. Recent advances in graphene family materials toxicity investigations. *J. Nanopart. Res.* **2012**, *14*, 1320–1340. [CrossRef] [PubMed]
44. Yuan, X.; Zhang, X.; Sun, L.; Wei, Y.; Wei, X. Cellular toxicity and immunological effects of carbon-based nanomaterials. *Part. Fibre Toxicol.* **2019**, *16*, 18–44. [CrossRef] [PubMed]
45. Buford, M.C.; Hamilton, R.F.; Holian, A. A comparison of dispersing media for various engineered carbon nanoparticles. *Part. Fibre Toxicol.* **2007**, *4*, 6. [CrossRef]
46. Rivankar, S. An overview of doxorubicin formulations in cancer therapy. *J. Cancer Res. Ther.* **2014**, *10*, 853–858. [CrossRef]
47. Shen, D.W.; Fojo, A.; Chin, J.E.; Roninson, B.; Richert, N.; Pastan, I.; Gottesman, M.M. Human multidrug-resistant cell lines: Increased mdr1 expression can precede gene amplification. *Science* **1986**, *232*, 643–645. [CrossRef]
48. Arora, H.C.; Jensen, M.P.; Yuan, Y.; Wu, A.; Vogt, S.; Paunesku, T.; Woloschak, G.E. Nanocarriers enhance Doxorubicin uptake in drug-resistant ovarian cancer cells. *Cancer Res.* **2012**, *72*, 769–778. [CrossRef]

49. Minotti, G.; Menna, P.; Salvatorelli, E.; Cairo, G.; Gianni, L. Anthracyclines: Molecular advances and pharmacologic developments in antitumor activity and cardiotoxicity. *Pharmacol. Rev.* **2004**, *56*, 185–229. [CrossRef]
50. Patil, R.R.; Guhagarkar, S.A.; Devarajan, P.V. Engineered nanocarriers of doxorubicin: A current update. *Crit. Rev. Ther. Drug Carr. Syst.* **2008**, *25*, 1–61. [CrossRef]
51. Tian, Z.; Yin, M.; Ma, H.; Zhu, L.; Shen, H.; Jia, N. Supramolecular assembly and antitumor activity of multiwalled carbon nanotube-camptothecin complexes. *J. Nanosci. Nanotechnol.* **2011**, *11*, 953–958. [CrossRef] [PubMed]
52. Scott, D.O.; Bindra, D.S.; Stella, V.J. Plasma pharmacokinetics of lactone and carboxylate forms of 20(S)-camptothecin in anesthetized rats. *Pharm. Res.* **1993**, *10*, 1451–1457. [CrossRef] [PubMed]
53. Hertzberg, R.P.; Caranfa, M.J.; Holden, K.G.; Jakas, D.R.; Gallagher, G.; Mattern, M.R.; Mong, S.M.; Bartus, J.O.; Johnson, R.K.; Kingsbury, W.D. Modification of the hydroxy lactone ring of camptothecin: Inhibition of mammalian topoisomerase I and biological activity. *J. Med. Chem.* **1989**, *32*, 715–720. [CrossRef]
54. Lu, J.; Liong, M.; Zink, J.I.; Tamanoi, F. Mesoporous silica nanoparticles as a delivery system for hydrophobic anticancer drugs. *Small* **2007**, *3*, 1341–1346. [CrossRef] [PubMed]
55. Pagura, C.; Barison, S.; Battiston, S.; Schiavon, M. Synthesis and characterization of single wall carbon nanohorns produced by direct vaporization of graphite. *TechConnect Briefs* **2010**, *1*, 289–291.
56. Hu, W.; Peng, C.; Lv, M.; Li, X.; Zhang, Y.; Chen, N.; Fan, C.; Huang, Q. Protein corona-mediated mitigation of cytotoxicity of graphene oxide. *ACS Nano* **2011**, *5*, 3693–3700. [CrossRef]
57. Xue, Z.; Wen, H.; Zhai, L.; Yu, Y.; Li, Y.; Yu, W.; Cheng, A.; Wang, C.; Kou, X. Antioxidant activity and anti-proliferative effect of a bioactive peptide from chickpea (Cicer arietinum L.). *Food Res. Int.* **2015**, *77*, 75–81. [CrossRef]
58. Rosenkranz, A.R.; Schmaldienst, S.; Stuhlmeier, K.M.; Chen, W.; Knapp, W.; Zlabinger, G.J. A microplate assay for the detection of oxidative products using 2′,7′-dichlorofluorescin-diacetate. *J. Immunol. Methods* **1992**, *156*, 39–45. [CrossRef]
59. Dehghankelishadi, P.; Dorkoosh, F.A. Pluronic based nano-delivery systems; Prospective warrior in war against cancer. *Nanomed. Res. J.* **2016**, *1*, 1–7. [CrossRef]
60. Zhang, W.; Gilstrap, K.; Wu, L.; Bahadur, R.; Moss, M.A.; Wang, Q.; Lu, X.; He, X. Synthesis and characterization of thermally responsive Pluronic F127-Chitosan nanocapsules for controlled release and intracellular delivery of small molecules. *ACS Nano* **2010**, *4*, 6747–6759. [CrossRef]
61. Wang, R.; Hughes, T.; Beck, S.; Vakil, S.; Li, S.; Pantano, P.; Draper, R.K. Generation of toxic degradation products by sonication of Pluronic® dispersants: Implications for nanotoxicity testing. *Nanotoxicology* **2013**, *7*, 1272–1281. [CrossRef] [PubMed]
62. Raval, A.; Pillai, S.S.; Bahadur, S.; Bahadur, P. Systematic characterization of Pluronic® micelles and their application for solubilization and in vitro release of some hydrophobic anticancer drugs. *J. Mol. Liq.* **2017**, *230*, 473–481. [CrossRef]
63. Aschberger, K.; Johnston, H.J.; Stone, V.; Aitken, R.J.; Hankin, S.M.; Peters, S.A.K.; Tran, C.L.; Christensen, F.M. Review of carbon nanotubes toxicity and exposure-appraisal of human health risk assessment based on open literature. *Crit. Rev. Toxicol.* **2010**, *40*, 759–790. [CrossRef] [PubMed]
64. Cui, D.; Tian, F.; Ozkan, C.S.; Wang, M.; Gao, H. Effect of single wall carbon nanotubes on human HEK293 cells. *Toxicol. Lett.* **2005**, *155*, 73–85. [CrossRef] [PubMed]
65. Jain, S.; Singh, S.R.; Pillai, S. Toxicity issues related to biomedical applications of carbon nanotubes. *J. Nanomed. Nanotechol.* **2012**, *3*, 5. [CrossRef]
66. Gökhan, M.M.; Scott-Budinger, G.R.; Alexander, A.G.; Daniela, U.; Saul, S.; Sergio, E.C.; George, F.A.; McCrimmon, D.R.; Igal, S.; Mark, C.H. Biocompatible nanoscale dispersion of single walled carbon nanotubes minimizes in vivo pulmonary toxicity. *Nano Lett.* **2010**, *10*, 1664–1670. [CrossRef]
67. Kobayashi, N.; Izumi, H.; Morimoto, Y. Review of toxicity studies of carbon nanotubes. *J. Occup. Health* **2017**, *59*, 394–407. [CrossRef]
68. Firme, C.P., III; Bandaru, P.R. Toxicity issues in the application of carbon nanotubes to biological systems. *Nanomed. NBM* **2010**, *6*, 245–256. [CrossRef]
69. Ajima, K.; Yudasaka, M.; Murakami, T.; Maigné, A.; Shiba, K.; Iijima, S. Carbon nanohorns as anticancer drug carriers. *Mol. Pharm.* **2005**, *2*, 475–480. [CrossRef]

70. Guerra, J.; Herrero, M.A.; Vázquez, E. Carbon nanohorns as alternative gene delivery vectors. *RSC Adv.* **2014**, *4*, 27315–27321. [CrossRef]
71. Azami, T.; Kasuya, D.; Yuge, R.; Yudasaka, M.; Iijima, S.; Yoshitake, T.; Kubo, Y. Large-scale production of single-wall carbon nanohorns with high purity. *J. Phys. Chem. C* **2008**, *112*, 1330–1334. [CrossRef]
72. Liu, J.; Cui, L.; Losic, D. Graphene and graphene oxide as new nanocarriers for drug delivery applications. *Acta Biomater.* **2013**, *9*, 9243–9257. [CrossRef] [PubMed]
73. Mohamadi, S.; Hamidi, M. Chapter 3—The new nanocarriers based on graphene and graphene oxide for drug delivery applications. In *Nanostructures for Drug Delivery*; Elsevier: Amsterdam, The Netherlands, 2017; pp. 107–147. [CrossRef]
74. Krüger, A.; Liang, Y.; Jarrea, G.; Stegk, J. Surface functionalisation of detonation diamond suitable for biological applications. *J. Mater. Chem.* **2006**, *16*, 2322–2328. [CrossRef]
75. Mochalin, V.N.; Shenderova, O.; Ho, D.; Gogotsi, Y. The properties and applications of nanodiamonds. *Nat. Nanotechnol.* **2012**, *7*, 11–23. [CrossRef]
76. Kaur, R.; Badea, I. Nanodiamonds as novel nanomaterials for biomedical applications: Drug delivery and imaging systems. *Int. J. Nanomed.* **2013**, *8*, 203–220. [CrossRef]
77. Zhu, Y.; Li, J.; Li, W.; Zhang, Y.; Yang, X.; Chen, N.; Sun, Y.; Zhao, Y.; Fan, C.; Huang, Q. The biocompatibility of nanodiamonds and their application in drug delivery systems. *Theranostics* **2012**, *2*, 302–312. [CrossRef]
78. Tu, Y.; Lv, M.; Xiu, P.; Huynh, T.; Zhang, M.; Castelli, M.; Liu, Z.; Huang, Q.; Fan, C.; Fang, H.; et al. Destructive extraction of phospholipids from Escherichia coli membranes by graphene nanosheets. *Nat. Nanotechnol.* **2013**, *8*, 594–601. [CrossRef]
79. Akhavan, O.; Ghaderi, E. Toxicity of graphene and graphene oxide nanowalls against bacteria. *ACS Nano* **2010**, *4*, 5731–5736. [CrossRef] [PubMed]
80. Zhang, J.; Sun, Q.; Bo, J.; Huang, R.; Zhang, M.; Xia, Z.; Ju, L.; Xiang, G. Single-walled carbon nanohorn (SWNH) aggregates inhibited proliferation of human liver cell lines and promoted apoptosis, especially for hepatoma cell lines. *Int. J. Nanomed.* **2014**, *9*, 759–773. [CrossRef]
81. Miyawaki, J.; Yudasaka, M.; Azami, T.; Kubo, Y.; Iijima, S. Toxicity of single-walled carbon nanohorns. *ACS Nano* **2008**, *2*, 213–226. [CrossRef] [PubMed]
82. He, B.; Shi, Y.; Liang, Y.; Yang, A.; Fan, Z.; Yuan, L.; Zou, X.; Chang, X.; Zhang, H.; Wang, X.; et al. Single-walled carbon-nanohorns improve biocompatibility over nanotubes by triggering less protein-initiated pyroptosis and apoptosis in macrophages. *Nat. Commun.* **2018**, *9*, 2393. [CrossRef] [PubMed]
83. Zhuang, W.R.; Wang, Y.; Cui, P.F.; Xing, L.; Lee, J.; Kim, D.; Jiang, H.L.; Oh, Y.K. Applications of π-π stacking interactions in the design of drug-delivery systems. *J. Control. Release* **2019**, *294*, 311–326. [CrossRef] [PubMed]
84. Hwang, Y.; Park, S.H.; Lee, J.W. Applications of functionalized carbon nanotubes for the therapy and diagnosis of cancer. *Polymers* **2017**, *9*, 13. [CrossRef] [PubMed]
85. Li, J.; Zhu, Y.; Li, W.; Zhang, X.; Peng, Y.; Huang, Q. Nanodiamonds as intracellular transporters of chemotherapeutic drug. *Biomaterials* **2010**, *31*, 8410–8418. [CrossRef]
86. Xi, G.; Robinson, E.; Mania-Farnell, B.; Fausto Vanin, E.; Shim, K.W.; Takao, T.; Allender, E.V.; Mayanil, C.S.; Soares, M.B.; Ho, D.; et al. Convection-enhanced delivery of nanodiamond drug delivery platforms for intracranial tumor treatment. *Nanomed. NBM* **2014**, *10*, 381–391. [CrossRef]
87. Welsher, K.; Liu, Z.; Sherlock, S.P.; Robinson, J.T.; Chen, Z.; Daranciang, D.; Dai, H. A route to brightly fluorescent carbon nanotubes for near-infrared imaging in mice. *Nat. Nanotechnol.* **2009**, *4*, 773–780. [CrossRef]
88. Lee, C.H.; Rajendran, R.; Jeong, M.S.; Ko, H.Y.; Joo, J.Y.; Cho, S.; Chang, Y.W.; Kim, S. Bioimaging of targeting cancers using aptamer-conjugated carbon nanodots. *Chem. Commun.* **2013**, *49*, 6543–6545. [CrossRef]
89. De la Zerda, A.; Zavaleta, C.; Keren, S.; Vaithilingam, S.; Bodapati, S.; Liu, Z.; Levi, J.; Smith, B.R.; Ma, T.-J.; Oralkan, O.; et al. Carbon nanotubes as photoacoustic molecular imaging agents in living mice. *Nat. Nanotechnol.* **2008**, *3*, 557–562. [CrossRef]
90. Kim, W.; Galanzha, E.I.; Shashkov, E.V.; Moon, H.-M.; Zharov, V.P. Golden carbon nanotubes as multimodal photoacoustic and photothermal high-contrast molecular agents. *Nat. Nanotechnol.* **2009**, *4*, 688–694. [CrossRef]

91. Nie, L.; Huang, P.; Li, W.; Yan, X.; Jin, A.; Wang, Z.; Tang, Y.; Wang, S.; Zhang, X.; Niu, G.; et al. Early-stage imaging of nanocarrier-enhanced chemotherapy response in living subjects by scalable photoacoustic microscopy. *ACS Nano* **2014**, *8*, 12141–12150. [CrossRef]
92. Sitharaman, B.; Shi, X.F.; Walboomers, X.F.; Liao, H.B.; Cuijpers, V.; Wilson, L.J.; Mikos, A.G.; Jansen, J.A. In vivo biocompatibility of ultra-short single-walled carbon nanotube/biodegradable polymer nanocomposites for bone tissue engineering. *Bone* **2008**, *43*, 362–370. [CrossRef] [PubMed]
93. Shi, X.F.; Sitharaman, B.; Pham, Q.P.; Liang, F.; Wu, K.; Billups, W.E.; Wilson, L.J.; Mikos, A.G. Fabrication of porous ultra-short single-walled carbon nanotube nanocomposite scaffolds for bone tissue engineering. *Biomaterials* **2007**, *28*, 4078–4090. [CrossRef] [PubMed]
94. Lee, H.J.; Park, J.; Yoon, O.J.; Kim, H.W.; Lee, D.Y.; Kim, D.H.; Lee, W.B.; Lee, N.E.; Bonventre, J.V.; Kim, S.S. Amine-modified single-walled carbon nanotubes protect neurons from injury in a rat stroke model. *Nat. Nanotechnol.* **2011**, *6*, 121–125. [CrossRef] [PubMed]
95. Keefer, E.W.; Botterman, B.R.; Romero, M.I.; Rossi, A.F.; Gross, G.W. Carbon nanotube coating improves neuronal recordings. *Nat. Nanotechnol.* **2008**, *3*, 434–439. [CrossRef] [PubMed]
96. Al-Jamal, K.T.; Nunes, A.; Methven, L.; Ali-Boucetta, H.; Li, S.; Toma, F.M.; Herrero, M.A.; Al-Jamal, W.T.; Ten Eikelder, H.M.M.; Foster, J.; et al. Degree of chemical functionalization of carbon nanotubes determines tissue distribution and excretion profile. *Angew. Chem. Int. Ed. Engl.* **2012**, *51*, 6389–6393. [CrossRef] [PubMed]
97. Bhattacharya, K.; Mukherjee, S.P.; Gallud, A.; Burkert, S.C.; Bistarelli, S.; Bellucci, S.; Bottini, M.; Star, A.; Fadeel, B. Biological interactions of carbon-based nanomaterials: From coronation to degradation. *Nanomed. NBM* **2016**, *12*, 333–351. [CrossRef]

© 2020 by the authors. Licensee MDPI, Basel, Switzerland. This article is an open access article distributed under the terms and conditions of the Creative Commons Attribution (CC BY) license (http://creativecommons.org/licenses/by/4.0/).

Article

Delivery of siRNA to Ewing Sarcoma Tumor Xenografted on Mice, Using Hydrogenated Detonation Nanodiamonds: Treatment Efficacy and Tissue Distribution

Sandra Claveau [1,2], Émilie Nehlig [3], Sébastien Garcia-Argote [3], Sophie Feuillastre [3], Grégory Pieters [3], Hugues A. Girard [4], Jean-Charles Arnault [4], François Treussart [1,5,*,†] and Jean-Rémi Bertrand [2,†]

1. LuMIn, CNRS, ENS Paris-Saclay, CentraleSupélec, Université Paris-Saclay, 91405 Orsay, France; sandra.claveau@live.fr
2. Vectorologie et Thérapeutiques Anticancéreuses, CNRS, Institut Gustave Roussy, Université Paris-Saclay, 94805 Villejuif, France; jean-remi.bertrand@gustaveroussy.fr
3. SCBM, Institut Joliot, CEA, Université Paris-Saclay, 91191 Gif-sur-Yvette, France; e.nehlig@gmail.com (É.N.); sebastien.garcia-argote@cea.fr (S.G.-A.); sophie.feuillastre@cea.fr (S.F.); gregory.pieters@cea.fr (G.P.)
4. Diamond Sensors Laboratory, Institut LIST, CEA, Université Paris-Saclay, 91191 Gif-sur-Yvette, France; hugues.girard@cea.fr (H.A.G.); jean-charles.arnault@cea.fr (J.-C.A.)
5. Institut d'Alembert, CNRS, ENS Paris-Saclay, Université Paris-Saclay, 91190 Gif-sur-Yvette, France
* Correspondence: francois.treussart@ens-paris-saclay.fr
† Co-senior authors.

Received: 7 February 2020; Accepted: 15 March 2020; Published: 19 March 2020

Abstract: Nanodiamonds of detonation origin are promising delivery agents of anti-cancer therapeutic compounds in a whole organism like mouse, owing to their versatile surface chemistry and ultra-small 5 nm average primary size compatible with natural elimination routes. However, to date, little is known about tissue distribution, elimination pathways and efficacy of nanodiamonds-based therapy in mice. In this report, we studied the capacity of cationic hydrogenated detonation nanodiamonds to carry active small interfering RNA (siRNA) in a mice model of Ewing sarcoma, a bone cancer of young adults due in the vast majority to the *EWS-FLI1* junction oncogene. Replacing hydrogen gas by its radioactive analog tritium gas led to the formation of labeled nanodiamonds and allowed us to investigate their distribution throughout mouse organs and their excretion in urine and feces. We also demonstrated that siRNA directed against *EWS-FLI1* inhibited this oncogene expression in tumor xenografted on mice. This work is a significant step to establish cationic hydrogenated detonation nanodiamond as an effective agent for in vivo delivery of active siRNA.

Keywords: nanodiamond; tritium; biodistribution; Ewing sarcoma; drug delivery; siRNA; nanomedicine

1. Introduction

The use of nanoparticles as vectors for drug delivery has been largely described by the scientific community during these past decades, with numerous applications [1,2]. Nanoparticles facilitate intracellular delivery and protection of the cargo against degradation, therefore they present many advantages for the vectorization of small nucleic acids such as small interfering RNA (siRNA). The latter are used to control gene expression by silencing targeted genes. Considering their intrinsic poor cellular penetration and low stability in biologicals medium [3,4], siRNA must be associated to an effective carrier. Different types of siRNA transporting agents have been reported either organic (e.g., liposomes, cationic polymers or dendrimers [5]), inorganic (e.g., metallic nanoparticles such as gold, iron, titanium [4]) or mineral like clay [6], silica nanoparticles [7] and nanodiamonds [8,9]. Most inorganic nanocarriers present a good-to-high chemical stability; they have a low toxicity at therapeutic dose of the drug:carrier conjugate and are able to deliver their cargo compounds into cells. Nevertheless, the high stability of inorganic nanoparticles goes with the fact that they are not biodegradable, and for pharmacological applications, this can be a limiting factor. Therefore, for the safe development of these particles, it is crucial to determine their possible elimination pathways after administration and study how they are distributed in the body.

Here, we report on the use of cationic detonation nanodiamonds (DND) for the delivery of siRNA directed against Ewing sarcoma (ES) junction oncogene *EWS-FLI1* to ES tumor xenografted on mice. Ewing sarcoma is a rare bone and soft tissue cancer [10] which is caused in the vast majority of cases by the formation of *EWS-FLI1* oncogene. To carry siRNA, DND surface needs to be modified to be able to bind negatively charged nucleic acids by electrostatic interactions. One strategy relies on cationic polymer coating of diamond surface, which is either done by electrostatic interaction [9,11] or by covalent grafting [12,13]. However, polymer coating may lead to the formation of large aggregates by crosslinking DND to other DND via polymer chains bridging. Another strategy, the one selected for this study, is based on the direct surface modification of DND using hydrogen gas combined with microwave plasma or with annealing method. Both hydrogenation methods were recently compared [14]. It was shown that after such reductive treatment, the hydrogenated DND (H-DND) can be dispersed in water, and acquire a positive surface charge characterized by a zeta potential of $\approx$+50 mV [15]. In a previous work, we described that H-DND can carry efficiently siRNA to ES cultured cells and promote specific targeted *EWS-FLI1* oncogene inhibition [16]. We want to extend this strategy to preclinical study now. Considering the very high chemical stability of diamond, these further investigations have to consider the risk of accumulation in the body after inoculation, as already described for larger nanodiamonds (around 50 nm diameter), produced by a different process than detonation [17]. However, owing to their small unitary size (3–8 nm) DND are good candidates for in vivo applications. Indeed, particles smaller than the filtration cutoff of kidney can be eliminated through urines, after glomerular passage, and since this limit for kidney is around 7–10 nm (depending of the molecule shape) [18,19], H-DND could be eliminated through this pathway. Indeed, Riojas et al. [20] showed that hydroxylated 7-nm sized DND further functionalized with radiolabeled amino groups are efficiently eliminated in urines if the solution is filtered before being intravenous injected.

In this work, we describe an original method that we developed to treat ES in vivo by injecting DND:siRNA complexes, and to determine the organ distribution of these DND. To this aim, we used DND labeled with tritium by annealing method [21]. We show that siRNA (i) can be loaded onto hydrogenated or tritiated DND (H-DND and T-DND, respectively), (ii) can efficiently inhibit *EWS-FLI1* in ES tumor model xenografted on mice and (iii) that the organ distribution and elimination of T-DND can be measured thanks to its radioactivity.

2. Materials and Methods

2.1. Preparation of Hydrogenated and Tritiated DND and Associated Suspensions

The procedures are detailed in reference [21]. DND (Reference: G02 grade, primary size 3–8 nm, from PlasmaChem GmbH, Berlin, Germany) were first manually milled in a mortar and then annealed using appropriate gas.

2.1.1. H-DND Aqueous Suspension

A mass of 30 mg of as-received and manually milled DND were placed in a quartz tube (3.5 mL) with an isolation valve, an in/out gas connection and connected to a cold trap. Vacuum was created and H_2 gas was loaded at 250 mbar pressure. The tube was placed in an oven, and connection was made with a trapping set-up. The oven was turned on during 1 h at 550 °C. The powder was then pumped for 30 min before disconnection and air exposure. Particles were dispersed in ultrapure water (resistivity: 18.2 MΩ.cm) and sonicated (Model UP400S, 300 W and 24 kHz, from Hielscher GmbH, Teltow, Germany) for 1 h under a cooling system. Aggregates of particles were then removed by centrifuging the suspension for 40 min (acceleration: 2400× g, speed: 4754 rpm) and the supernatant was collected, forming the stock H-DND aqueous suspension. The final concentration was calculated by measuring the mass of particles after drying a calibrated volume of the supernatant.

2.1.2. Mixture of T-DND and H-DND (Later Denoted T-DND to Simplify) Aqueous Suspension

As-received manually milled DND were first pre-oxidized to remove native C-H groups to achieve a more quantitative control of tritium added to the surface [21]. To this aim, 100 mg was placed in an alumina crucible under air for 1 h 30 min at 550 °C. Then, 30 mg of these pre-oxidized DND were annealed for 4 h at 550 °C, using the same process that for H-DND with 10% T_2 and 90% H_2 gas mixture (in order to obtain the desired activity). To remove all the labile tritium atoms, the treated powder was poured into methanol (4 mL) and the solvent was evaporated. The operation was repeated twice and the powder was then stored dried under a nitrogen atmosphere. Quantification of the tritium incorporation was assessed by measuring the activity (using liquid scintillation counting) in the combustion gas after the total combustion of T-DND in air (3 h at 600 °C). We measured a total specific activity of 13 mCi.mg^{-1}.

The synthesized T-DND (4 mg) were dispersed in ultrapure water (3 mL) and sonicated (3 mm conical microprobe Vibracell 75043, 750 W, 28% amplitude, Sonics & Materials, Inc., Newton, CT, USA) for 1 h under a cooling system. To remove highly aggregated particles, the suspensions were centrifuged for 40 min (2400× g, 4754 rpm) followed by supernatant separation. A liquid scintillation counting was performed on the supernatant, yielding a specific activity of 18 mCi/mL.

To reduce the high specific activity of the suspension, T-DND from the supernatant (10 μL) were mixed with a solution of H-DND (2 mg/mL, 2.1 mL). The later was prepared following Section 2.1.1 method with a small adjustment: in order to be consistent with the tritium treatment, this H-DND suspension was also prepared from pre-oxidized DND treated for 4 h under H_2 at 550 °C. The final DND solution was split in 3 sealed vials (0.7 mL each). The total activity measured by liquid scintillation counting of this final solution was 97 μCi/mL.

2.1.3. Additional Pre-Injection Washing of T-DND Solution.

Before injection to mice, the T-DND solution was centrifuged (acceleration 10,600× g for 3 h at 10 °C) in order to eliminate residual labile tritium atoms by exchange with water. Washed T-DND could then be stably suspended in water, and only 2% of the total initial tritium radioactivity was lost in the supernatant (see Supplementary Table S1).

2.2. Hydrodynamic Size and Electrophoretic Mobility Characterizations DND Solution

Hydrodynamic size and zeta potential of H-DND and DND:siRNA complexes in solution were measured by dynamic light scattering (DLS) using a NanoBrook 90Plus PALS (Brookhaven Instruments, Holtsville, NY, USA) in 1 cm thick cuvette in deionized water. Hydrodynamic sizes are inferred from the scattered intensity autocorrelation function. The latter was then analyzed with the non-negative constrained least squared method [22], which is one of the methods of reference to infer the size from polydisperse suspensions. The size values reported correspond to the dominant population. For radioactive nanodiamonds, measurements were performed in a sealed cuvette.

2.3. siRNA Sequences and Binding siRNA to Hydrogenated or Tritiated DND and siRNA Binding Capacity Assay

2.3.1. siRNA Sequences

siRNA was purchased from Kaneka Eurogentec S.A. (Seraing, Belgium). The sequence complementary to the *EWS/FLI1* fusion oncogene (siAS) is: sense strands 5′-GCAGCAGAACCCUUC UUAUd(GA)-3′ and antisense strand 5′-AUAAGAAGGGUUCUGCUGCd(CC)-3′. The control irrelevant sequence (siCt) is: sense strand 5′-CGUUACCAUCGAGGAUCCAd(AA)-3′ and the antisense strand 5′-UGGAUCCUCGAUGGUAACGd(CT)-3′.

2.3.2. Binding siRNA to Hydrogenated or Tritiated DND and siRNA Binding Capacity Assay

siRNA complexation to cationic DND was performed by slowly dropping a siRNA solution to cationic DND solutions placed in a sonication bath (Ultrasonic cleaner, VWR International S.A.S., Fontenay-sous-Bois, France) at its maximum power during 10 min, maintaining room temperature in the bath. The measurement of hydrodynamic size and zeta potential were performed afterwards. The determination of the binding capacity of siRNA to nanodiamonds was performed by mixing to a fixed concentration of siRNA (0.3 µg/mL) an increasing concentration of H-DND from 0 to 600 µg/mL in 100 mM NaCl, 10 mM HEPES pH 7.2 buffer, in a fixed 60 µL final volume. After centrifugation (16,000× *g* at 10 °C for 15 min), non-complexed free siRNA concentrations were measured on 30 µL of the supernatants to which an equal volume of 1 µg/mL ethidium bromide (EtBr, Sigma-Aldrich S.a.r.l, Saint-Quentin Fallavier, France) was added. The mixtures were placed into a 96-wells plate, then analyzed with a fluorescence plate reader (Glomax Multi+, Promega, Charbonnières-les-Bains, France) set at 525 nm excitation and 580–640 nm bandpass detection wavelengths, to infer the free siRNA amount. The results are presented as the fraction of the fluorescence intensity relative to the one of the total amounts of siRNA before adding DND. Experiments were realized in triplicate.

2.4. Measurement of EWS-FLI1 Inhibition in Cultured Cells by RT-qPCR

One day before treatment, 3×10^5 human Ewing sarcoma cells A673 were seeded in 12 wells-plate in DMEM medium (Gibco, Life Technologies S.A.S., Courtaboeuf, France) containing 10% fetal calf serum (Gibco, Life Technologies S.A.S., Courtaboeuf, France) and 1% Penicillin, streptomycin solution (Gibco, Life Technologies S.A.S., Courtaboeuf, France) and then incubated at 37 °C, with 5% CO_2 and 95% hygrometry. Then the medium was discarded and replaced by 500 µL of 75 nM siRNA bound to H-DND at a mass ratio of 5:1, 25:1 or 50:1 (H-DND:siRNA) in DMEM medium containing 10% fetal calf serum and 1% penicillin and streptomycin solution, for 24 h at 37 °C, 5% CO_2 and 95% hygrometry. Comparatively, same conditions are used with 75 nM of siAS bounded to Lipofectamine 2000 (Life Technologies S.A.S., Courtaboeuf, France) added to cells in serum containing medium. Then, the medium was discarded, and the cells were lysed by 400 µL of Trizol solution (Invitrogen) and collected in Eppendorf tubes. Total RNA were extracted by adding 60 µL of chloroform:isoamyl alcohol (49:1). After centrifugation at 13,000 rpm for 15 min at 10 °C, 150 µL of the aqueous phase was precipitated by adding 150 µL isopropanol for 15 min at room temperature followed by centrifugation for 15 min at 13,000 rpm. Pellets were washed twice with 70% ethanol solution and dried. Plellets were then solubilized with water (10 µL) containing 0.5 U/µL of RNasin (Promega, Charbonnières-les-Bains, France) and RNA was quantified with a spectrophotometer (NanoDrop™, Life Technologies S.A.S., Courtaboeuf, France). The cDNA was prepared by heating 1.5 µg of RNA in 12.5 µL of water at 75 °C for 5 min and with 2 µL of 50 µg/mL random primer (Promega). Then, 4 µL of 5X M-MLV buffer (Promega), 0.5 µL dNTP 20 mM (Promega), 0.5 µL RNasin 40 U/µL (Promega) and 0.5 µL M-MLV Reverse Transcriptase 200 U/µL (Promega) were added and incubation was performed for 15 min at 25 °C followed by 1 h at 42 °C. *EWS-FLI1* mRNA expression is performed by qPCR on 5 µL of 1/20 diluted cDNA, 0.4 µL of each primer at 10 mM concentration, 4.2 µL deionized water and 10 µL of 2X KAPA SYBR® FAST Master Mix (Sigma-Aldrich S.a.r.l, Saint-Quentin Fallavier, France).

PCR was performed for 40 cycles in fast mode on a StepOnePlus™ system (Applied Biosystems, Life Technologies S.A.S., Courtaboeuf, France). *EWS-FLI1* gene was amplified with *EWS*-forward primer: 5′-AGC AGT TAC TCT CAG CAG AAC ACC-3′ and *FLI1*-reverse: 5′-CCA GGA TCT GAT ACG GAT CTG GCC-3′. As a control, we used GAPDH gene with forward primer: 5′-CAA GGT CAT CCATGA CAA CTT TG-3′, and reverse primer: 5′-GTC CAC CAC CCT GTT GCT GTA G-3′. We normalized the number PCR cycle threshold C_t for the target sequence to the one for GAPDH control gene. Experiments are performed in triplicate.

2.5. In vivo Experiments

2.5.1. Animal Experimentation

Animal experiments were performed in accordance with the ethical project submitted and approved by the ethical committee N°26 regulating the animal facility at Gustave Roussy Institute (Villejuif, France) and under national agreement N°2013-062-01223.

2.5.2. Biodistribution of Nanodiamonds in Mice

We injected 100 µL of PBS buffer containing 3×10^6 A673 cells in the right flank of male nude mice. When the tumors appeared, we injected in the tail veins of the mice 100 µL of water for injectable preparation containing either siRNA (5 mg/kg) alone, T-DND (25 mg/kg) alone or siRNA complexed to T-DND (5 mg/kg siRNA mixed to 25 mg/kg T-DND, i.e., mass ratio: 5:1). We then placed the mice in metabolic cages for 4 h or 24 h. At these times, 3 mice of each condition were humanely sacrificed, the tissues were withdrawn and their weights are determined. About 0.1 g of each tissue is sampled and solubilized with 1 mL of Solvable™ (PerkinElmer, Courtaboeuf, France) heated during 3 h at 55 °C. After the transfer of the solubilized tissues in scintillation tube, solutions were clarified by adding 100 µL of 30% hydrogen peroxide (Sigma-Aldrich S.a.r.l, Saint-Quentin Fallavier, France). Then 10 mL of Ultima Gold (PerkinElmer, Courtaboeuf, France) were added before the radioactivity measurements were performed during 1 min with 1409 DSA scintillation counter (Wallac/PerkinElmer, Courtaboeuf, France). The results are expressed as deionized counts per minutes per unit mass of tissue (cpm/g) or as the percentage of the injected dose to the whole organs. All urines from cabinet containing animal group receiving a same treatment (in these conditions, measurement is a global statistical value) were collected. Experiments were carried out in triplicates for each condition. For a healthy mouse, the glomerular filtration rate of a compound intravenously injected is about 7 µL/min per g, leading to an almost full elimination after 3 h [23]. Therefore, at our observation times of 4 h and 24 h, we should be able to detect T-DND in urines, if they could be eliminated through this pathway.

2.5.3. Measurement of *EWS-FLI1* Inhibition in Nude Mice

EWS-FLI1 expression in tumor was measured after intravenous injection in the tail vein of nude mice of 100 µL of antisense or control siRNA 5 mg/kg bound to H-DND at mass ratio of 5:1 (H-DND:siRNA) or free H-DND for 24 h. Then a fragment of the tumor was sampled and placed in a tube containing 400 µL of Trizol solution (Invitrogen, Life Technologies S.A.S., Courtaboeuf, France). A ceramic ball was added and tissue homogenization was performed for 30 s in TissuLyser II (Quiagen Paris, Courtaboeuf, France). Finally, *EWS-FLI1* expression was quantified by RT-qPCR as previously described.

2.6. Statistical Analysis

Data are presented as means and either standard deviation or standard error on the mean. Statistical significance was evaluated using GraphPad Instat 3.10 software (GraphPad Sofware, San Diego, CA, USA). For all the comparisons between two conditions, we used the non-parametric Wilcoxon–Mann–Whitney two sample rank test (two-tailed *p* value provided), while for the comparison between more than three conditions we applied a Kruskal–Wallis test, followed by Dunn's Multiple Comparisons Test. One star (*) corresponds to $p < 0.05$, ** to $p < 0.01$ and *** to $p < 10^{-3}$.

3. Results

3.1. Characterization of the Hydrogenated and Tritiated Nanodiamond Suspensions

After hydrogenation or tritiation by annealing using molecular gases (see methods and [21]), we suspended the treated DND powder in distilled water by sonication. We measured hydrodynamical diameters of DND clusters ranging from about 60 nm to about 90 nm for H-DND and T-DND colloidal suspensions, respectively (Supplementary Table S2). These results are in agreement with previous colloidal studies on hydrogenated DND [21,24] and also with the transmission electron microscopy observations we reported in Ref. [16] (see Figure 4a in this reference), showing H-DND aggregates of ≈50 nm in size at the cell membrane.

3.2. Loading Capacity of Nanodiamonds for siRNA Binding

We quantified the loading capacity of H-DND for siRNA binding. To this aim, we measured by fluorescence (see Materials and Methods), for a fixed amount of siRNA, the remaining free unbound siRNA for increasing amounts of hydrogenated nanodiamonds. The results are presented in Figure 1 for two conditions: one in which H-DND and siRNA were simply mixed, the other in which the mixing was done under sonication, both for the same duration. For these two conditions, the complete fixation of siRNA occurred at the same mass ratio $m_{H\text{-}DND}/m_{siRNA}$ of 10:1. These data suggest the sonication assistance leads to a faster adsorption of siRNA onto H-DND, resulting from the larger thermal agitation brought by ultrasound waves.

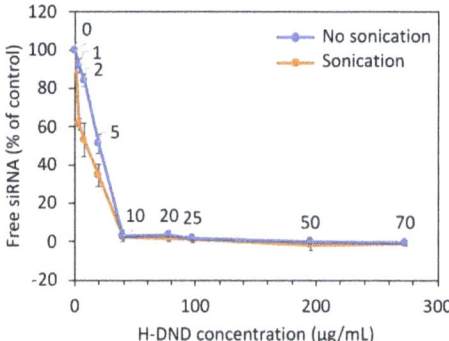

Figure 1. Binding of siRNA to hydrogenated detonation nanodiamonds (H-DND). Free siRNA as function of increasing quantity of H-DND (H-DND-22 samples) added to a fixed initial quantity of siRNA. The results are presented as the percentage of initial siRNA. Orange/blue with or without (respectively) sonication during the siRNA to H-DND complexation. Experiments were performed in triplicates. Number over each point represent the mass ratio $m_{H\text{-}DND}/m_{siRNA}$. The statistical comparison tests between sonicated and non-sonicated conditions yield $p = 0.021$ (*) for mass ratio 1:1 to 5:1 and non-significant differences starting from mass ratio 10:1 to higher ones.

3.3. Effect of siRNA Binding to Nanodiamonds on the Hydrodynamic Size and Zeta Potential of Complexes

Figure 2 displays the variation of the hydrodynamic size and zeta potential of H-DND:siRNA complexes upon increase of the DND:siRNA mass ratio. The experiment was carried out under two conditions: with or without sonication during complexation. We observed that for mass ratios lower than 10 the particles hydrodynamic size is around 80 nm (Figure 2a). Then, the size increased strongly for mass ratio between 10 and 40 to return to the initial size for mass ratio higher than 40. We conclude that strong aggregation occurred only for mass ratios between 10 and 40.

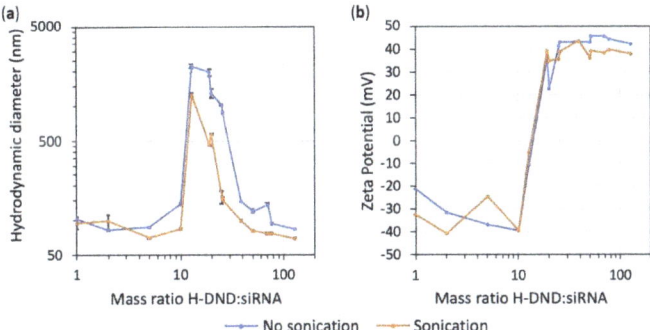

Figure 2. Variation in hydrodynamic size (in intensity) and zeta potential of siRNA bound to increasing mass ratio of H-DND (H-DND-22 sample). Complexes were prepared with or without sonication. (**a**) Size measurement by dynamic light scattering (DLS). (**b**) Zeta potential determination.

Looking at the zeta potentials of each solution (Figure 2b), we observed that for mass ratios lower than 10, the surface charge is negative (around −30 mV) corresponding to an excess of siRNA covering nanodiamonds. For mass ratios between 10 and 40, a surface charge inversion occurs with a zeta potential close to +40 mV for a mass ratio of 40. In between, the low zeta potential of the complexes promotes their aggregation, as observed on Figure 2a. For mass ratios higher than 40, the positive zeta potential reflects an excess of hydrogenated nanodiamonds. Sonication have no effect on surface charge but promote smaller aggregates, as observed on the DLS profiles (Figure 2a).

Moreover, considering that we inject the H-DND:siRNA conjugate in the blood, we could have some concern regarding its aggregation in such a complex environment. However, it is well established (see for example S. Hamelaar et al. [25]) that the serum favors the dispersion of electrostatically-charged nanoparticles when they are dispersed in a culture medium. In order to confirm these data with our specific conjugate, we formed a solution of H-DND:siRNA at the same mass ratio of 5:1 used in the in vivo experiment, and measured its hydrodynamic diameter in DMEM culture medium supplemented with 10% Fetal Bovine Serum. We found a diameter of 35 nm, a value even smaller than the diameter of 90 nm of the same conjugate in pure water (see Table S2), which is a strong indication that injecting intravenously an aqueous suspension of H-DND:siRNA (mass ratio 5:1), will rather induce a small deagglomeration than an aggregation in the blood circulation.

3.4. Inhibition of EWS-FLI1 on Ewing Sarcoma Cultured Cells

We first studied the efficacy of H-DND prepared by annealing under hydrogen gas, as vector for siRNA delivery to human Ewing sarcoma A673 cells. We used different mass ratios between H-DND and siRNA, from 5:1 to 50:1. We observed on Figure 3 that 35% inhibition is obtained with these cationic H-DND for mass ratios higher than 25. In comparison, commercially available cationic liposomes Lipofectamine 2000 used in similar conditions in the presence of serum during transfection show a lower efficacy with only 18% inhibition. It is to note that in recommended conditions (transfection in serum free medium) Lipofectamine 2000 yields 70% inhibition on this cell model [9]. The efficacy of siRNA transfection by H-DND in serum containing medium is crucial for further applications in animals.

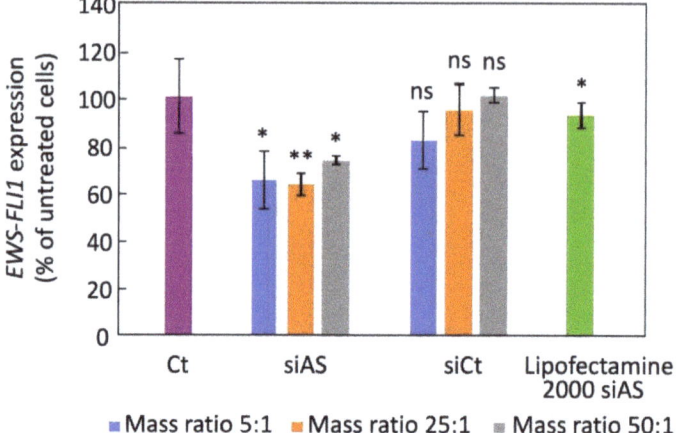

Figure 3. Inhibition of *EWS-FLI1* expression in A673 Ewing sarcoma cultured cells by siRNA vectorized by H-DND (sample H-DND-24) or Lipofactamine 2000. Cells are treated during 24 h in DMEM medium containing 10% of fetal calf serum by 75 nM siRNA. The H-DND:siRNA mass ratio was 5:1, 25:1 or 50:1. siAS: antisense siRNA directed against *EWS-FLI1* oncogene; siCt: control irrelevant siRNA sequence (see methods). Experiments were performed in triplicate. Comparisons were all done relative to the "Ct" condition using the Wilcoxon–Mann–Whitney two sample rank test, that yielded $p_{5:1} = 0.018$, $p_{25:1} = 0.009$; $p_{50:1} = 0.036$, $p_{lipo} = 0.019$; *ns*: non-significant.

3.5. Inhibition of EWS-FLI1 Expression in Mice

We then evaluated the efficacy of siRNA vectorized by T-DND to inhibit *EWS-FLI1* in tumor xenografted on mice. We produced *EWS-FLI1* tumor model by grafting A673 cells in the flank of nude mice. We used tritiated DND to conduct both the inhibition and biodistribution studies using the same mice. We selected the mass ratio of 5:1, first of all to ensure that all T-DND are covered by siRNA. Our hypothesis was that such a configuration would favor the detachment of siRNA molecules which interact only by a portion of their total length with the T-DND surface. Moreover, the mass ratio of 5:1 also ensures that the injected suspension is not too viscous, since at the given necessary dose of siRNA and volume of solution to be injected, the concentration of T-DND remains low enough to have a small impact on the viscosity of the aqueous suspension. Finally, high dose of cationic vector may be toxic as observed in cultured cells. For all these reasons, we favored the smallest H-DND concentration, corresponding to the 5:1 mass ratio.

We injected the T-DND radiolabels in the tail vein of mice, using either uncomplexed T-DND or with siRNA complexed to them. We did not include a free siRNA group because we already established that free siRNA is not able to inhibit *EWS-FLI1* oncogene [26]. We considered 6 to 8 nude mice per condition, which were placed in metabolic cage. After 4 h or 24 h, mice were humanly sacrificed. For the efficacy study, we only considered the 24 h group. Using RNA extracted from tumors, we then quantified *EWS-FLI1* expression by RT-qPCR. The normal level of *EWS-FLI1* expression is provided by mice treated by non-complexed T-DND. We also used an irrelevant siRNA complexed with T-DND as a specificity control. We observed on Figure 4 that the irrelevant T-DND:siRNA treatment has no effect. The *EWS-FLI1* antisense siRNA complexed to T-DND inhibits the *EWS-FLI1* expression by about 50%. These results confirm that H/T-DND:siRNA is able to inhibit *EWS-FLI1* in this Ewing sarcoma tumor model grafted on nude mice.

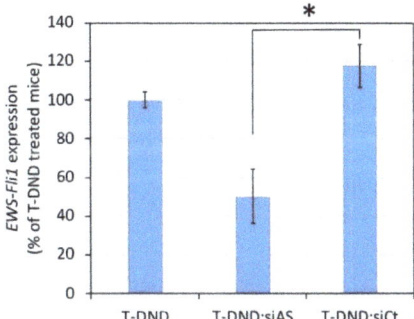

Figure 4. Inhibition of *EWS-FLI1* expression in tumor xenografted on mice treated for 24 h by siRNA vectorized by tritiated DND (T-DND) at a mass ratio of 5:1 (T-DND:siRNA). The tumor was formed from A673 Ewing sarcoma cells grafted in the right flank of the mice (n = 6 to 8 animals per condition). T-DND:siRNA was intravenously injected in the mouse tail vein. RNA was extracted from the tumor of mice sacrificed 24 h after treatment, and RT-qPCR was performed using GAPDH as housekeeping gene. Standard error on the mean are indicated. siAS:siRNA against *EWS-FLI1* oncogene; siCt: irrelevant. A significant difference between T-DND:siAS and T-DND-siCt is observed (*) according to the p_{Dunn} value <0.05, inferred from a Kruskal–Wallis test (p = 0.0094) followed by a Dunn multiple comparisons test.

3.6. Biodistribution of Nanodiamonds in Nude Mice

For an efficient use of H-DND produced by annealing for siRNA delivery to Ewing sarcoma tumor xenografted on mice, it is important to determine their tissue distribution and elimination, which is made possible by radioactivity measurement using their radioactive analogs T-DND. The different tissues of the mice of the efficacy study (Section 3.5) were collected, homogenized and then the radioactivity was determined as presented in Figure 5. We also collected urines and feces at the same time points. We observed that T-DND accumulated mainly in the liver, lung, spleen, kidney and also the heart for T-DND:siRNA. There are no significant changes between 4 h and 24 h in the quantity of T-DND found in the different tissues. For kidney, spleen, lung and heart, T-DND complexed with siRNA accumulated more than uncomplexed T-DND. The radioactivity is recovered in all tested tissues at a lower level. We did not observe high accumulation in the tumor. Nevertheless, the dose measured in the tumor is four time larger than the one in the blood.

One of the major questions for the use of mineral origin nanoparticles for biomedical applications is how they may be eliminated after injection in animals. The radioactive fraction recovered in urines was 0.15% of the injected dose per mouse. Unfortunately, after centrifugation of urines, we found that all radioactivity was recovered in the supernatant. This indicates that no T-DND was eliminated through the kidney pathway. The very small detected radioactivity is probably the one of water consecutive to exchange of H with labile T. Note that 4 h after injection in the kidney we detected about 0.25% of the radioactive T-DND:siRNA injected dose. This can be due either to the smallest T-DND being filtrated by the glomeruli but then reabsorbed by the tubules and/or to T-DND aggregates not being able to cross the glomeruli. This last hypothesis is in agreement with the fact that uncomplexed T-DND accumulated about twice less in kidney than the one associated to siRNA, which present a higher aggregation state. Another possible elimination pathway is by the feces. Thanks to the metabolism cage, we estimated that the total collected amount of feces after 24 h had a radioactivity representing about 0.19% for H-DND:siRNA and 0.13% for H-DND of the injected doses per mouse (Figure 5b), which is about 6 times larger than the one after 4 h (0.03% and 0.02% for H-DND:siRNA and H-DND, respectively), indicating a linear elimination with time, in perfect agreement with the constant value of radioactivity in feces sampled in the rectum at 4 h and 24 h (Figure 5a). Indeed, since the majority of DND are captured by liver cells, their transfer to the intestine lumen by the bile may be the main way for T-DND elimination.

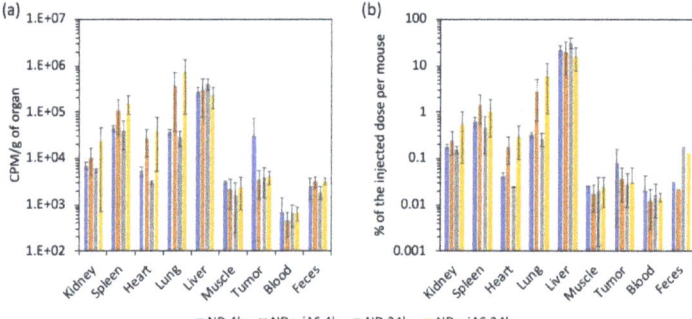

Figure 5. Mouse organ distribution of uncomplexed T-DND or complexed with siRNA at mass ratio 5:1 of T-DND:siRNA. The total radioactivity of each tissue is measured after 4 h and 24 h for 3 animals per series. Error bars: standard deviations. (**a**) Radioactivity presented as count per minutes (cpm) per gram of tissue or feces. The latter corresponds only to the fraction collected in the rectum of each animal. (**b**) Radioactivity from each full organ, presented as the percentage of the total injected dose. Feces were collected for all the animals simultaneously in the same metabolism cage, but the total amount, including non-expelled feces is not known which is why we could not include feces data in B. ND + siAS: DND complexed with siRNA sequence antisense for *EWS-FLI1* oncogene. The differences between error bars for the different conditions are due to (i) the variability between animal groups, and (ii) to the fact that the T-DND quantifications were done only on portions of ≈100 mg of each organ, and that we cannot exclude an inhomogeneous distribution of T-DND within the organs.

4. Discussion

In a previous work, we had demonstrated that cationic hydrogenated detonation nanodiamonds can deliver active siRNA to Ewing Sarcoma cell in culture [16]. Here, we wanted to establish if small hydrogenated DND are able to deliver active siRNA in vivo to mice model of Ewing sarcoma. In order to investigate simultaneously their biodistribution, we produced hydrogenated and tritiated detonation nanodiamonds via their exposure to a mixture of H_2 and T_2 molecular gases at 550 °C under controlled pressure. The characterization in size and surface charge was performed by DLS measurement and zeta potential determination (Table S2). We observed that H-DND are cationic with a zeta potential of ≈+45 mV and have a hydrodynamic size of ≈60 nm. H-DND positive charge favored electrostatic attachment of siRNA onto the nanoparticles and the measured DLS size larger than the 7-nm primary size indicated some aggregation. This size difference is partly due to the measurement methods, as unitary diamond size is the one of its hard core as observed by high-resolution transmission electron microscopy, whereas DLS yields the hydrodynamic size, that is larger because it takes into account the water dipole molecule attached to the core and moving with it under Brownian motion. Our data indicate that H-DND are present in the aqueous suspension in the form of aggregates of an average number of eight primary 7 nm sized particles.

The binding capacity of H-DND was studied by mixing to a constant quantity of siRNA an increasing quantity of H-DND and the full binding of siRNA was obtained for 10 times excess of DND (Figure 1). Hydrodynamic size and zeta potential variations during the titration process of siRNA with H-DND (Figure 2a) indicated that for mass ratio around this point of saturation (in the 10 to ≈40 mass ratio range) the size of H-DND:siRNA complexes increased to form large aggregates. The sonication during siRNA and H-DND association process decreased the size of the aggregates but did not prevent it. For mass ratio smaller than 10 or larger than ≈40 the nanoparticle hydrodynamic sizes were close to the ones of uncomplexed H-DND (<100 nm, Table S2). Furthermore, the slightly smaller size of the H-DND:siRNA complex under sonication indicates its high colloidal stability. This strong binding capacity of H-DND presents advantages for animal applications to prevent desorption of the cargo before complexes reach their target cells. Finally, during the titration process, the zeta potential decreased from +50 mV for an excess of H-DND (mass ratios larger than 20) to a negatively charged complexes (zeta potential ≈-30 mV) for mass ratios smaller than 10 (Figure 2b). The sonication did not change the mass ratio transition threshold. This observation is consistent with the aggregation happening once all nanodiamonds have lost their charges due to the complexation with siRNA. It is worth to note that the global charge of the complexes could be tuned to negative (mass ratios smaller than 10) or to positive (mass ratios larger than 20), resulting in the formation of complexes that may interact differently with cell membrane and lead to different siRNA delivery efficacy [27].

In the prospect of using H-DND for anticancer gene delivery in mice, we first validated the delivery of these cationic DND by their capacity to transport a siRNA cargo directed against *EWS-FLI1* oncogene in cultured A673 human Ewing sarcoma cells and by detecting a gene inhibition efficacy. We observed that H-DND vectorized siRNA inhibited *EWS-FLI1* expression by 30–40% (Figure 3). In comparison with previous results [16], the inhibition of *EWS-FLI1* expression by siRNA transported by DND hydrogenated via annealing is lower than the one obtained with plasma hydrogenated DND vectors. This may be due to the slightly larger aggregation of DND hydrogenated via annealing, leading to a lower internalization efficiency [28].

One key question regarding the use of non-biodegradable nanoparticles for drug delivery in animal is to determine whether they are eliminated, through which pathways, and where in the body the non-eliminated fraction accumulates. It was shown that "large" ND of 50 nm primary size reside in animal tissues a few weeks after injection [17]. However, "small" ND of size compatible with kidney elimination may overcome this limitation. Indeed, nanoparticles smaller than 6–8 nm have been already reported to be filtered at the glomeruli level [18,19]. In this respect, detonation ND with primary size included between 3 and 8 nm and even smaller [29] are good candidates for such elimination. For this study, we prepared radioactive tritiated DND to be able to trace these ultra-small particles after injection to mice. We observed that the liver, spleen and lung contained much more radioactive amount than the other tissues (Figure 5). Moreover, T-DND carrying siRNA accumulated more in kidney, spleen, heart and lung, with a slow increase from 4 to 24 h. Unfortunately, we observed a limited accumulation of T-DND in the tumor compared to other tissues such as muscles, heart or kidney. This indicated that Enhanced Permeability and Retention (EPR) [30] effect did not occur for these complexes or/and that the tumor had no aberrant vascularization responsible for EPR effect. Nevertheless, T-DND:siRNA quantity was stable in the tumor between 4 h to 24 h which is the time range needed to induce the silencing effect. During the first 24 h, the blood radioactivity did almost not decrease, which indicated that DND were only very slowly eliminated from the circulation.

The main possible elimination pathways are urines and feces. Radioactivity from urines could come from both T-DND or tritiated water. The later could originate from a fraction of tritium desorbing from T-DND and exchanging with hydrogen in water. Using ultracentrifugation, we measured this labile tritium fraction to be 5% (see Table S1). In order to reduce the amount of free tritium in the injected solution, all T-DND:siRNA complexes were prepared from T-DND purified by this ultracentrifugation process. However, even after such purification, there is a remaining fraction of 2% free tritium. This value provides a lower bound of T-DND detection sensitivity in urines. Urines were collected globally for each group of mice and their radioactivity was measured in 1 mL of solution and in 1 mL of ultra-centrifugated supernatant. We observed the same values of radioactivity before and after centrifugation (Table S1) within a margin of error larger than the detection sensitivity lower bound. Hence, the radioactivity found in urines most likely comes from the residual free tritium present in the injected solution and not from T-DND. We therefore consider that T-DND were not eliminated in urines, or only very slowly. The possible reasons for the absence of T-DND in urine are (i) the formation of aggregates too large to be filtrated or/and (ii) ultrasmall DND reabsorbed by kidney tubules, the first one (i) being the most likely, considering previous observations of Rojas et al. [20]

who showed that membrane-filtrated DND are eliminated in urines, much more efficiently than the aggregates dominant in the original suspension.

We then explored the other possible elimination pathway, by collecting the feces from the rectum and we measured their radioactivity. We observed a high radioactive signal in the feces. The amount of radioactivity in the feces fraction is larger than the one of free tritium in the injected volume, according to the labile tritium release measurements (Table S1), therefore we shall conclude that DND are present in the feces which therefore constitute one of the elimination pathways. It is likely that DND are eliminated though the bile by liver filtration. Since the bile goes into the intestinal lumens, DND are then incorporated to the feces. This result is of high importance because it states that ultrasmall DND can be partly eliminated from mouse body after intravenous administration.

Furthermore, if we consider the radioactivity balance between injected doses per mouse and the total amount collected, only about 25% of the dose is recovered in the sampled fractions. The 75% of not-measured radioactive T-DND are localized in the carcass containing skin, not removed muscles, bones and so on, and probably a part has also been eliminated through other routes. Indeed, the elimination pathways are multiple in animals. Some authors propose expectoration as an alternative way of elimination. Considering that DND also accumulated into the lung (Figure 5), they may be engulfed by macrophages which further direct them into the alveoli where they may be finally expulsed through the pharynx by the mucociliary system [31]. Further investigations would be necessary to evaluate the expectorated fraction.

Finally, regarding the silencing efficacy of *EWS-FLI1* in Ewing sarcoma cells xenografted tumor by siRNA delivered by T-DND, we observed a 50% inhibition by the antisense siRNA compared to irrelevant siRNA. The latter yielded a similar effect than the T-DND alone, i.e., no inhibition. This result indicates that despite a low accumulation of T-DND in the tumor, the siRNA is delivered to the cancer cells where they silence the targeted oncogene. Note that although the in vitro inhibition efficacy is similar to the in vivo one, one cannot generally extrapolate in vitro results to in vivo ones which justifies our study. Finally, in order to treat the mouse and reduce the tumor size, we could combine a conventional anticancer treatment like vincristine to the H-DND-siRNA one, as we showed in our previous in vitro study [16].

Overall, our study represents a significant step towards the use of ultra-small solid nanoparticles which are able to deliver efficiently active siRNA to tumor cell in animals and are also eliminated from their body subsequently.

Supplementary Materials: The following are available online at http://www.mdpi.com/2079-4991/10/3/553/s1, Table S1: Determination of free tritium in nanodiamond suspension and in mice urine after injection; Table S2: Size measurement of hydrogenated and tritiated DND conjugated or not to siRNA, in water and culture medium supplemented with serum.

Author Contributions: J.-R.B. and F.T. conceived and designed the experiments; S.C. and J.-R.B.; performed the experiments; S.C., F.T. and J.-R.B. analyzed the data; É.N., S.G.-A., H.A.G., J.-C.A., S.F. and G.P. prepared the hydrogenated and tritiated nanodiamond solutions; J.-R.B. and F.T. wrote the article and did the figures. All authors have read and agreed to the published version of the manuscript.

Funding: This research was funded by French National Research Agency ANR through the ERANET Euronanomed 2 program, grant number ANR-14-ENM2-0002.

Acknowledgments: We thank the animal experimentation platform from Gustave Roussy Institute for technical assistance and in particular Mélanie Polrot for her expertise in animal manipulation.

Conflicts of Interest: The authors declare no conflict of interest.

References

1. Ho, B.N.; Pfeffer, C.M.; Singh, A.T.K. Update on nanotechnology-based drug delivery systems in cancer treatment. *Anticancer Res.* **2017**, *37*, 5975–5981. [CrossRef]
2. Shi, J.; Kantoff, P.W.; Wooster, R.; Farokhzad, O.C. Cancer nanomedicine: Progress, challenges and opportunities. *Nat. Rev. Cancer* **2017**, *17*, 20–37. [CrossRef]

3. Barata, P.; Sood, A.K.; Hong, D.S. RNA-targeted therapeutics in cancer clinical trials: Current status and future directions. *Cancer Treat. Rev.* **2016**, *50*, 35–47. [CrossRef]
4. Marquez, A.R.; Madu, C.O.; Lu, Y. An overview of various carriers for siRNA delivery. *Oncomedicine* **2018**, *3*, 48–58. [CrossRef]
5. Jiang, W.; Von Roemeling, C.A.; Chen, Y.; Qie, Y.; Liu, X.; Chen, J.; Kim, B.Y.S. Designing nanomedicine for immuno-oncology. *Nat. Biomed. Eng.* **2017**, *1*, 1–11. [CrossRef]
6. Castro-Smirnov, F.A.; Ayache, J.; Bertrand, J.-R.; Dardillac, E.; Le Cam, E.; Piétrement, O.; Aranda, P.; Ruiz-Hitzky, E.; Lopez, B.S. Cellular uptake pathways of sepiolite nanofibers and DNA transfection improvement. *Sci. Rep.* **2017**, *7*, 5586. [CrossRef] [PubMed]
7. Oshiro Junior, J.; Paiva Abuçafy, M.; Berbel Manaia, E.; Lallo da Silva, B.; Chiari-Andréo, B.; Aparecida Chiavacci, L.; Oshiro Junior, J.A.; Paiva Abuçafy, M.; Berbel Manaia, E.; Lallo da Silva, B.; et al. Drug delivery systems obtained from silica based organic-inorganic hybrids. *Polymers* **2016**, *8*, 91. [CrossRef]
8. Ho, D.; Wang, C.-H.K.; Chow, E.K.-H. Nanodiamonds: The intersection of nanotechnology, drug development, and personalized medicine. *Sci. Adv.* **2015**, *1*, e1500439. [CrossRef]
9. Alhaddad, A.; Adam, M.-P.; Botsoa, J.; Dantelle, G.; Perruchas, S.; Gacoin, T.; Mansuy, C.; Lavielle, S.; Malvy, C.; Treussart, F.; et al. Nanodiamond as a vector for siRNA delivery to ewing sarcoma cells. *Small* **2011**, *7*, 3087–3095. [CrossRef]
10. Cidre-Aranaz, F.; Alonso, J. EWS/FLI1 target genes and therapeutic opportunities in ewing sarcoma. *Front. Oncol.* **2015**, *5*, 162. [CrossRef]
11. Chen, M.; Zhang, X.-Q.; Man, H.B.; Lam, R.; Chow, E.K.; Ho, D. Nanodiamond vectors functionalized with polyethylenimine for siRNA delivery. *J. Phys. Chem. Lett.* **2010**, *1*, 3167–3171. [CrossRef]
12. Neburkova, J.; Vavra, J.; Cigler, P. Coating nanodiamonds with biocompatible shells for applications in biology and medicine. *Curr. Opin. Solid State Mater. Sci.* **2017**, *21*, 43–53. [CrossRef]
13. Sotoma, S.; Hsieh, F.-J.; Chang, H.-C. Single-step metal-free grafting of cationic polymer brushes on fluorescent nanodiamonds. *Materials* **2018**, *11*, 1479. [CrossRef] [PubMed]
14. Arnault, J.-C.; Girard, H.A. Hydrogenated nanodiamonds: Synthesis and surface properties. *Curr. Opin. Solid State Mater. Sci.* **2017**, *21*, 10–16. [CrossRef]
15. Ginés, L.; Mandal, S.; Ashek-I-Ahmed, A.-I.-A.; Cheng, C.-L.; Sow, M.; Williams, O.A.; Gines, L.; Mandal, S.; Ahmed, A.-I.; Cheng, C.-L.; et al. Positive zeta potential of nanodiamonds. *Nanoscale* **2017**, *9*, 12549–12555. [CrossRef]
16. Bertrand, J.-R.; Pioche-Durieu, C.; Ayala, J.; Petit, T.; Girard, H.A.; Malvy, C.P.; Le Cam, E.; Treussart, F.; Arnault, J.-C. Plasma hydrogenated cationic detonation nanodiamonds efficiently deliver to human cells in culture functional siRNA targeting the Ewing sarcoma junction oncogene. *Biomaterials* **2015**, *45*, 93–98. [CrossRef]
17. Yuan, Y.; Chen, Y.; Liu, J.-H.; Wang, H.; Liu, Y.; Lui, J.-H. Biodistribution and fate of nanodiamonds in vivo. *Diam. Relat. Mater.* **2009**, *18*, 95–100. [CrossRef]
18. Longmire, M.; Choyke, P.L.; Kobayashi, H. Clearance properties of nano-sized particles and molecules as imaging agents: Considerations and caveats. *Nanomedicine* **2008**, *3*, 703–717. [CrossRef]
19. Liu, J.; Yu, M.; Zhou, C.; Zheng, J. Renal clearable inorganic nanoparticles: A new frontier of bionanotechnology. *Mater. Today* **2013**, *16*, 477–486. [CrossRef]
20. Rojas, S.; Gispert, J.D.; Martin, R.; Abad, S.; Menchon, C.; Pareto, D.; Victor, V.M.; Alvaro, M.; Garcia, H.; Herance, J.R. Biodistribution of amino-functionalized diamond nanoparticles. In vivo studies based on 18 F radionuclide emission. *ACS Nano* **2011**, *5*, 5552–5559. [CrossRef]
21. Nehlig, E.; Garcia-Argote, S.; Feuillastre, S.; Moskura, M.; Charpentier, T.; Schleguel, M.; Girard, H.A.; Arnault, J.-C.; Pieters, G. Using hydrogen isotope incorporation as a tool to unravel the surfaces of hydrogen-treated nanodiamonds. *Nanoscale* **2019**, *11*, 8027–8036. [CrossRef] [PubMed]
22. Brown, W. *Dynamic Light Scattering: The Method and Some Applications*; Clarendon Press: Oxford, UK, 1993; ISBN 9780198539421.
23. Sasaki, Y.; Iwama, R.; Sato, T.; Heishima, K.; Shimamura, S.; Ichijo, T.; Satoh, H.; Furuhama, K. Estimation of glomerular filtration rate in conscious mice using a simplified equation. *Physiol. Rep.* **2014**, *2*, e12135. [CrossRef]

24. Petit, T.; Girard, H.A.; Trouvé, A.; Batonneau-Gener, I.; Bergonzo, P.; Arnault, J.-C. Surface transfer doping can mediate both colloidal stability and self-assembly of nanodiamonds. *Nanoscale* **2013**, *5*, 8958–8962. [CrossRef] [PubMed]
25. Hemelaar, S.R.; Nagl, A.; Bigot, F.; Rodríguez-García, M.M.; de Vries, M.P.; Chipaux, M.; Schirhagl, R. The interaction of fluorescent nanodiamond probes with cellular media. *Microchim. Acta* **2017**. [CrossRef] [PubMed]
26. Toub, N.; Bertrand, J.-R.; Tamaddon, A.; Elhamess, H.; Hillaireau, H.; Maksimenko, A.; Maccario, J.; Malvy, C.; Fattal, E.; Couvreur, P. Efficacy of siRNA nanocapsules targeted against the *EWS-FLI1* oncogene in Ewing sarcoma. *Pharm. Res.* **2006**, *23*, 892–900. [CrossRef]
27. Dante, S.; Petrelli, A.; Petrini, E.M.; Marotta, R.; Maccione, A.; Alabastri, A.; Quarta, A.; De Donato, F.; Ravasenga, T.; Sathya, A.; et al. Selective targeting of neurons with inorganic nanoparticles: Revealing the crucial role of nanoparticle surface charge. *ACS Nano* **2017**, *11*, 6630–6640. [CrossRef]
28. Zhang, S.; Gao, H.; Bao, G. Physical principles of nanoparticle cellular endocytosis. *ACS Nano* **2015**, *9*, 8655–8671. [CrossRef]
29. Stehlik, S.; Varga, M.; Ledinsky, M.; Miliaieva, D.; Kozak, H.; Skakalova, V.; Mangler, C.; Pennycook, T.J.; Meyer, J.C.; Kromka, A.; et al. High-yield fabrication and properties of 1.4 nm nanodiamonds with narrow size distribution. *Sci. Rep.* **2016**, *6*, 38419. [CrossRef]
30. Greish, K. *Enhanced Permeability and Retention (EPR) Effect for Anticancer Nanomedicine Drug Targeting*; Humana Press: Totowa, NJ, USA, 2010; pp. 25–37.
31. Yuan, Y.; Wang, X.; Jia, G.; Liu, J.-H.; Wang, T.; Gu, Y.; Yang, S.-T.; Zhen, S.; Wang, H.; Liu, Y. Pulmonary toxicity and translocation of nanodiamonds in mice. *Diam. Relat. Mater.* **2010**, *19*, 291–299. [CrossRef]

 © 2020 by the authors. Licensee MDPI, Basel, Switzerland. This article is an open access article distributed under the terms and conditions of the Creative Commons Attribution (CC BY) license (http://creativecommons.org/licenses/by/4.0/).

Article

Carbon Dot Nanoparticles Exert Inhibitory Effects on Human Platelets and Reduce Mortality in Mice with Acute Pulmonary Thromboembolism

Tzu-Yin Lee [1], Thanasekaran Jayakumar [1], Pounraj Thanasekaran [2], King-Chuen Lin [3,4], Hui-Min Chen [5], Pitchaimani Veerakumar [3,4,6],* and Joen-Rong Sheu [1,*]

[1] Graduate Institute of Medical Sciences, College of Medicine, Taipei Medical University, Taipei 110, Taiwan; d119103001@tmu.edu.tw (T.-Y.L.); tjaya_2002@yahoo.co.in (T.J.)
[2] Department of Chemistry, Fu Jen Catholic University, New Taipei City 242, Taiwan; ptsekaran@gmail.com
[3] Department of Chemistry, National Taiwan University, Taipei 10617, Taiwan; kclin@ntu.edu.tw
[4] Institute of Atomic and Molecular Sciences, Academia Sinica, Taipei 10617, Taiwan
[5] Department of Anatomy and Cell Biology, School of Medicine, College of Medicine, Taipei Medical University, Taipei 110, Taiwan; chm7805@tmu.edu.tw
[6] School of Chemistry, Madhurai Kamaraj University, Madhurai 625021, India
* Correspondence: spveerakumar@gmail.com (P.V.); sheujr@tmu.edu.tw (J.-R.S.); Tel.: +886-2-3366-1162 (P.V.); +886-2-27361661 (ext. 3199) (J.-R.S.)

Received: 26 May 2020; Accepted: 23 June 2020; Published: 28 June 2020

Abstract: The inhibition of platelet activation is considered a potential therapeutic strategy for the treatment of arterial thrombotic diseases; therefore, maintaining platelets in their inactive state has garnered much attention. In recent years, nanoparticles have emerged as important players in modern medicine, but potential interactions between them and platelets remain to be extensively investigated. Herein, we synthesized a new type of carbon dot (CDOT) nanoparticle and investigated its potential as a new antiplatelet agent. This nanoparticle exerted a potent inhibitory effect in collagen-stimulated human platelet aggregation. Further, it did not induce cytotoxic effects, as evidenced in a lactate dehydrogenase assay, and inhibited collagen-activated protein kinase C (PKC) activation and Akt (protein kinase B), c-Jun N-terminal kinase (JNK), and p38 mitogen-activated protein kinase (MAPK) phosphorylation. The bleeding time, a major side-effect of using antiplatelet agents, was unaffected in CDOT-treated mice. Moreover, our CDOT could reduce mortality in mice with ADP-induced acute pulmonary thromboembolism. Overall, CDOT is effective against platelet activation in vitro via reduction of the phospholipase C/PKC cascade, consequently suppressing the activation of MAPK. Accordingly, this study affords the validation that CDOT has the potential to serve as a therapeutic agent for the treatment of arterial thromboembolic disorders

Keywords: nanoparticles; carbon dots; platelet aggregation; arterial thrombosis; signaling molecules; bleeding disorder

1. Introduction

Platelet activation has been associated with several thrombotic diseases. While it plays a vital role in regulating hemorrhagic events, hyperactivity can lead to a range of complications. In general, patients with cardio- and cerebrovascular ailments are found to have more reactive platelets than normal, healthy individuals. Thrombotic diseases pose a severe threat to humans as they may elicit significant injury and even lead to death. Several studies have recommended that intravenous heparin and tissue plasminogen activators are effective for treatment [1,2]; nevertheless, these are unsafe and may lead to severe bleeding and problems associated with reocclusion and reinfarction [3]. The inhibitors of antiplatelet drugs, such as the P_2Y_{12} receptor, integrin $\alpha_{IIb}\beta_3$, cyclooxygenase, and phosphodiesterase, are also widely used; however, they have serious limitations. Further, phosphatidylinositol 3-kinase inhibitors have been proposed as potential antithrombotic agents [4], but they also are associated with some major restrictions for use as drugs.

Nanoparticles can be defined as any particulate materials that range from 1 to 100 nm in size in at least one dimension [5], and they are ubiquitously distributed in the environment. In fact, humans are often exposed to airborne nanoparticles [6]. Their size can be manipulated to facilitate their passage across biological membranes and affect cell physiology [7]. Berry et al. [8] reported the substantial accumulation of nanoparticles in platelets in pulmonary capillaries and anticipated that there might be a predisposing factor for platelet aggregation and microthrombi formation. Nanoparticles are not proposed for systemic use as they can interfere with platelet function and increase the risk of cardiovascular diseases and vascular thrombosis [9]. However, some types of nanoparticles have been developed for therapeutic purposes that can target the injured vascular site to mimic platelet function [10] or enhance blood clotting [11]. Nevertheless, their undesirable, antiaggregating properties are of a significant concern in nanomedicine, impeding their widespread application in the clinical setting.

Carbon dots (CDOTs) have become the most important type of nanoparticles, considering their favorable biological properties. They are obtained from natural carbon sources and their average diameter is < 10 nm [12]. These nanoparticles have even been acquired from organic substances and are constant in water media, which is tremendously noteworthy from a biological point of view [13], especially in drug delivery, bioimaging, optical imaging, and biosensing due to their high biocompatibility [14]. A study reported that, in comparison to CDOTs, the application of noncarbon quantum dots has not received adequate attention, as they are associated with severe health and environmental concerns [15]. Yan et al. [16] reported the antihemorrhagic effects of novel water-soluble carbon quantum dots, and their results indicated the explicit hemostatic effect of these nanoparticles. CDOTs, isolated from egg yolk oil, demonstrated a hemostatic effect in mice via the stimulating intrinsic blood coagulation and fibrinogen systems [17]. Another relevant study showed that CDOTs from the Phellodendri Cortex carbonisatus considerably reduced bleeding time and coagulation parameters and significantly increased platelets without inducing toxicity when administered in mice [18]. Mariangela Fedel has recently reviewed the hemocompatibility of carbon nanostructures [19]. However, in general, the antiplatelet aggregating effects of CDOTs have not been extensively explored. Therefore, in this study, we investigated the antiplatelet and antithrombotic effects of a new type of CDOT in human platelets and mice, respectively.

2. Materials and Methods

2.1. Reagents

Collagen (type I), 9, 11-dideoxy-11α, 9α-epoxymethanoprostaglandin (U46619), and thrombin were purchased from Chrono-Log Corporation (Havertown, PA, USA). Anti-phospho-p38 mitogen-activated protein kinase (MAPK) (Thr180/Tyr182), anti-phospho-c-Jun N-terminal kinase (JNK) (Thr183/Tyr185), anti-phospho-(Ser) protein kinase C (PKC) substrate, anti-JNK polyclonal antibodies (pAb), and anti-p38 MAPK and anti-Akt monoclonal antibodies (mAb) were purchased from Cell Signaling Technology (Beverly, MA, USA). Anti-phospho-Akt (Ser473) pAb was purchased from Biorbyt (Cambridge, UK), and anti-pleckstrin (p47) pAb was purchased from Gene Tex (Irvine, CA, USA). Hybond-P polyvinylidene difluoride (PVDF) membranes, enhanced chemiluminescence (ECL) Western blotting detection reagent, and the analysis system were purchased from GE Healthcare Life Sciences (Buckinghamshire, UK). Horseradish peroxidase-conjugated goat anti-rabbit and anti-mouse immunoglobulin G antibodies were obtained from Jackson ImmunoResearch Laboratories (West Grove, PA, USA).

2.2. Preparation of CDOTs

Fresh garlic (*Allium sativum*) cloves were purchased from a local market in Taiwan, which were then peeled, crushed, and suspended in ultrapure water. This suspension was vigorously stirred for 1 h at 40 °C. The extract was filtered twice to remove insoluble materials and then freeze-dried. The obtained powder was stored at −20 °C until required. For CDOT synthesis, 100 mg of the garlic extract powder was dissolved in 3 mL water and poly (diallyldimethylammonium chloride) mixture (1/0.5, *v/v*). The clear transparent solution obtained was heated in a domestic microwave oven for 5 min at 600 W and then cooled to ambient temperature (25 °C). The obtained yellow-brown solution was diluted with ultrapure water and dialyzed against water for 2–3 h through a dialysis membrane.

2.3. Characterization of the Synthesized CDOTs

Crystallographic information pertaining to the CDOTs was collected using an analytical X-ray diffractometer (X'Pert PRO, Malvern, Worcestershire, UK) using Cu Kα radiation (λ = 0.1541 nm). A Fourier transform infrared (FT-IR) spectrometer (Bruker IFS28, Billerica, MA, USA) in the range of 4000–400 cm^{-1} was used for the characterization of functional groups on the surface of the CDOTs, with an average of 21 scans. The sample was prepared as pellets using KBr (IR grade). Ultraviolet–visible (UV–vis) spectra were documented using a Thermo Scientific Evolution 220 spectrophotometer (Waltham, MA, USA), whereas fluorescence spectral measurements were taken using a PerkinElmer LS-45 spectrometer (Waltham, MA, USA). The morphological information on the prepared CDOTs was obtained through field-emission transmission electron microscopy (FE-TEM, JEOL JEM-2100F, Akishima, Tokyo, Japan).

2.4. Preparation of Washed Human Platelets and Lactate Dehydrogenase (LDH) Release Assay

This study was performed in accordance with the Declaration of Helsinki, and the Institutional Review Board of Taipei Medical University approved all protocols (IRB: N201612050). All volunteers provided informed consent before they participated in this study. Anticoagulated human blood with acid–citrate–dextrose (1:9) was collected from healthy human volunteers who had not eaten any drugs within a time of two weeks prior to the analysis. The method described by Sheu et al. [20] was used for preparing human platelet suspensions. The platelets were suspended in Tyrode's solution, and calcium chloride was then added, with the final concentration of Ca^{2+} being 1 mM.

The cytotoxicity of the CDOTs was evaluated using an LDH release assay. Washed platelets (3.6×10^8 cells/mL) were pretreated with 50-500 μM CDOTs or a solvent control (PBS; phosphate-buffered saline) for 20 min at 37 °C and then centrifuged at 5000 g for 5 min. The supernatant obtained was used for the assay. Briefly, 10 μL of the supernatant was placed on a Fuji Dri-Chem slide (LDH-PIII) (Tokyo, Japan), and the absorbance was measured at 540 nm using a UV–vis spectrophotometer (UV-160; Shimadzu, Japan). The LDH activity of 1% Triton X-100-treated washed platelets indicated 100% LDH release.

2.5. Platelet Aggregation

Platelet aggregation was monitored using a lumi-aggregometer (Helena Laboratories, Beaumont, TX, USA), as previously described [20]. The platelet suspension (3.6×10^8 cells/mL) was preincubated with various CDOT concentrations (25–120 μM) or an isovolumetric solvent control (PBS) for 3 min before the agonists were added. The reaction was permitted to continue for at least 10 min and the level of aggregation was calculated in light transmission units. The amplitude and slope of platelet aggregation were automatically calculated using the aggregometer.

2.6. Western Blotting

Washed platelets (1.2×10^9 cells/mL) were preincubated with the CDOTs (65 μM and 90 μM) for 3 min before treating them with collagen to induce platelet activation. A lysis buffer (200 μL) was used for platelet resuspension after the reaction was complete. Proteins (80 µg) from the supernatants were separated using 12% SDS-PAGE and electrophoretically transferred to PVDF membranes (Bio-Rad, Hercules, CA, USA). The membranes were blocked with 5% BSA in Tris-buffered saline (10 mM Tris-base, 100 mM NaCl, and 0.01% Tween 20) for 1 h and probed with various primary antibodies, followed by incubation with horseradish peroxidase-labeled anti-rabbit or anti-mouse immunoglobulin G antibodies for 1 h. Antibody-bound proteins on the membranes were detected using an ECL system and quantified using Bio-profil Biolight (version V2000.01; Vilber Lourmat, Marne-la-Vallée, France).

2.7. Tail Bleeding Time in Mice

ICR mice (20–25 g, 5–6 weeks old, male) were obtained from BioLasco (Taipei, Taiwan). All procedures and protocols were approved by the Affidavit of Approval of Animal Use Protocol, Taipei Medical University (LAC-2018-0360). The bleeding time was measured after 10 min of intravenous administration of 1 mg/kg CDOTs or PBS (control). The tail of anesthetized mice was cut 3 mm from the end and then directly immersed in normal saline at 37 °C. The bleeding time was recorded until no sign of bleeding was observed for at least 10 s.

2.8. ADP-Induced Acute Pulmonary Thromboembolism in Mice

According to our previous method, we used ADP to induce acute pulmonary thromboembolism in mice [21]. A fixed dose of the CDOTs (1 mg/kg) or PBS was intravenously injected, and after 10 min, ADP (0.7 mg/g) was injected into the tail vein. The lungs were then removed and placed in 4% formalin, and paraffin-embedded sections were stained with hematoxylin–eosin and then photographed using ScanScope CS (Leica Biosystems, Wetzlar, Germany). The mortality rate was recorded in all animal groups within 1 h of the injection.

2.9. Statistical Analysis

Data are expressed as mean ± standard error of the mean (SEM), and convoyed by the number of observations (n). n represents the number of experiments, and each experiment was performed using different blood donors. Statistical significances were analyzed for the in vivo experiments using unpaired Student's t test. One-way analysis of variance (ANOVA) was implemented to determine variations between the experimental groups and, if the analysis exhibited a significant difference, they were compared using the Student–Newman–Keuls test. $p < 0.05$ indicated statistical significance.

3. Results

3.1. Characterization of the CDOTs

3.1.1. X-ray Diffraction Analysis

The X-ray diffraction (XRD) pattern revealed that one diffraction peak at 2θ of 23.6° corresponded to disordered carbon atoms and the (002) graphite lattice, as shown in Figure 1A, and this finding was consistent with that previously reported for CDOTs [22].

3.1.2. FT–IR

The FT–IR spectrum observed for the garlic clove is rather similar to that of the CDOTs, indicating that the functional groups were, indeed, successfully provided the garlic clove, as illustrated in Figure 1B. The broad absorption band centering at 3427 cm^{-1} should be associated with the O–H stretching vibration mode of the hydroxyl functional groups in the garlic clove. The weak bands at 2940 and 1413 cm^{-1} confirm the presence of CH_2 groups, whereas the bands at 921 and 1568 cm^{-1} revealed the presence of oxygen-containing functional groups. The peaks at approximately 2944 and 1405 cm^{-1} were assigned to the C–H and C–N stretching vibration modes, and the absorption at 680 cm^{-1} could be ascribed to the C–S group [23]. Consequently, the as-prepared CDOTs were mainly composed of different functional groups on their surface, which is favorable for sustainable applications in biology.

3.1.3. UV–vis Spectroscopy

The UV–vis absorption spectra of the CDOTs, as shown by the blue line in Figure 1C, showed a comparable absorption band ranging from 200 to 600 nm, concordant with an earlier study on N-doped CDOTs produced by Wu et al. [24]. The CDOTs water solution produced solid blue light under UV irradiation of 365 nm, as shown by the right inset in Figure 1C. The CDOTs exhibited very strong FL in the range of 380–600 nm, with the maximum peak at around 446 nm, as shown by the red line in Figure 1C.

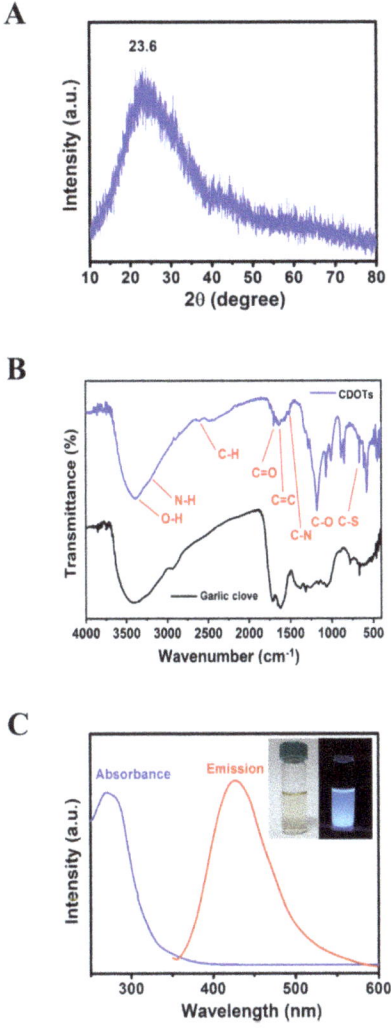

Figure 1. Characterization of the synthesized CDOTs. (**A**) X-ray diffraction (XRD), (**B**) Fourier transform infrared (FT–IR) spectra, and (**C**) UV–vis absorption spectra, as described in the Materials and Methods section.

3.2. LDH Assay and FE-TEM

Herein, we explored the probable toxic effects of the synthesized CDOTs on platelets by observing the release of cytosolic LDH. The CDOTs (50 µM and 100 µM) did not provoke any substantial discharge of LDH from platelet cytosol, even at concentrations of up to 200 µM, as shown in Figure 2A. Thus, they evidently did not disturb platelet membrane integrity or induce cytotoxicity at concentrations as high as 200 µM. A slight increase was observed at a higher concentration of 500 µM. LDH activities measured from the 1% Triton X-100-treated platelets were regarded as 100% release.

The morphological features and average particle sizes of the CDOTs are shown in Figure 2B. The synthesized CDOTs had a crystalline structure and were well distributed in water without aggregation. Furthermore, they were round in shape with a normal diameter of 3 nm [25].

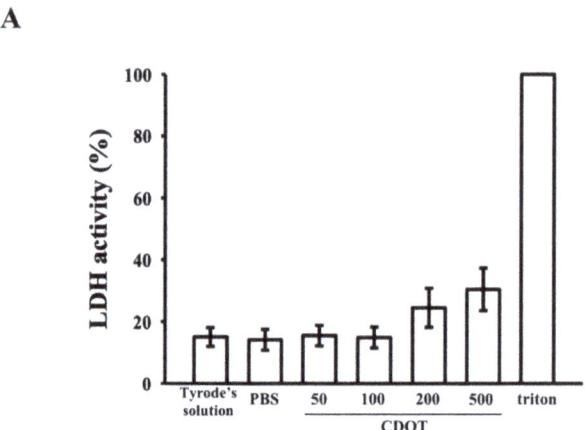

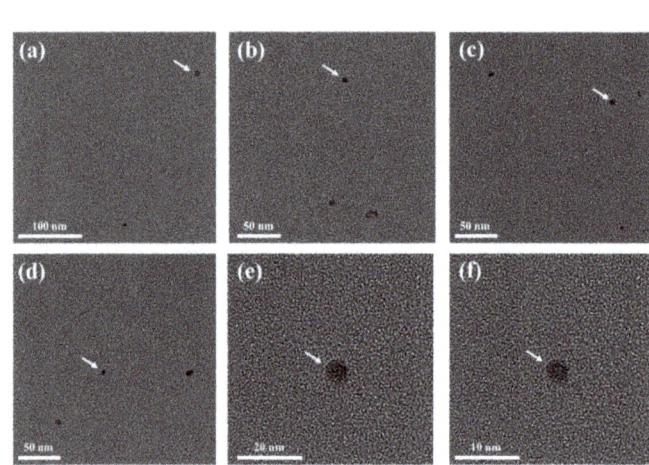

Figure 2. Cytotoxicity and morphology of the CDOTs. (**A**) Washed platelets (3.6×10^8 cells/mL) were preincubated with PBS (control) or the synthesized CDOTs (50, 100, 200 and 500 µM) for 20 min, and a 10 µL suspension of the supernatant was deposited on a Fuji Dri-Chem slide (LDH-PIII). (**Ba–f**) Field-emission transmission electron microscopic images. The arrows indicate sizes and morphologies of CDOTs. Values represent mean ± SEM ($n = 6$).

3.3. Inhibition of Platelet Aggregation Stimulated by Collagen

The CDOTs led to concentration-dependent (25–120 μM) inhibition of platelet aggregation induced by collagen (1 μg/mL), as shown in Figure 3A,B, but not by U46619 (1 μM), a prostaglandin endoperoxide (thromboxane A_2 receptor agonist), or thrombin (0.01 U/mL), even at higher concentrations of 120 μM, as shown in Figure 3C,D. Almost full inhibition was observed at 90 μM in collagen stimulated aggregation, as shown in Figure 3B. As a result, the IC_{50} (65 μM) and maximal concentration (90 μM) of the CDOTs were chosen to observe the potential inhibitory mechanisms in collagen-activated platelets. The CDOTs suppressed maximal platelet aggregation, stimulated by collagen, but not by U46619 and thrombin, as shown in Figure 4A, whereas the slopes of platelet aggregation revealed that the CDOTs also significantly reduced the lag time induced by these agonists, respectively, as shown in Figure 4B.

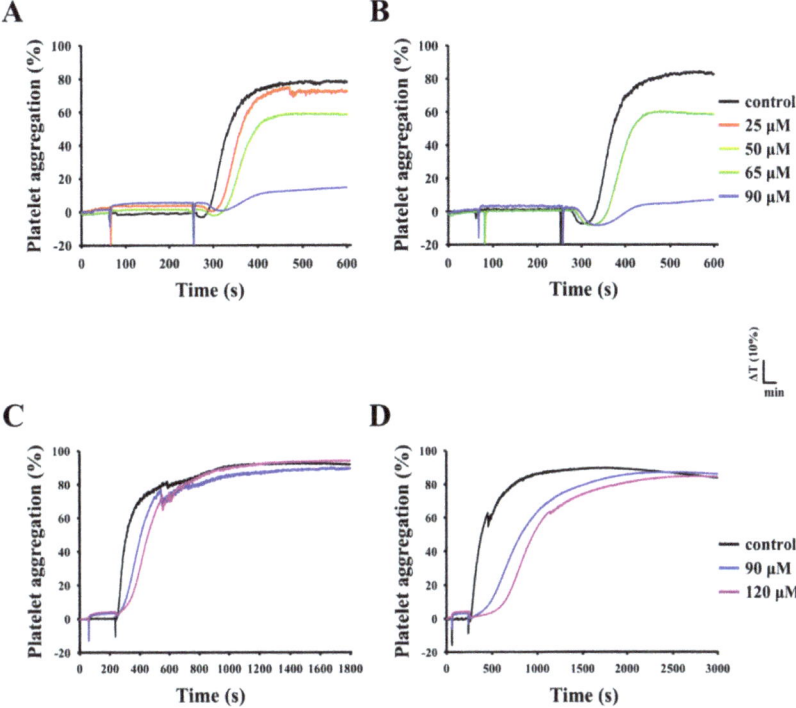

Figure 3. Inhibitory effects of the CDOTs on human platelet aggregation. Washed platelets (3.6 × 10^8 cells/mL) were preincubated with PBS (control) or the synthesized CDOTs (25-120 μM) and subsequently treated with (**A,B**) 1 μg/mL collagen, (**C**) 1 μM U46619, and (**D**) 0.01 U/mL thrombin to induce platelet aggregation. The IC50 and maximal inhibitory concentrations are shown in **B**. The inhibitory profiles (**A–D**) are representative examples of four similar experiments. The delayed lag phase of platelet aggregation noticed in CDOT-pretreated platelets stimulated by either U46619 (**C**) or thrombin (**D**).

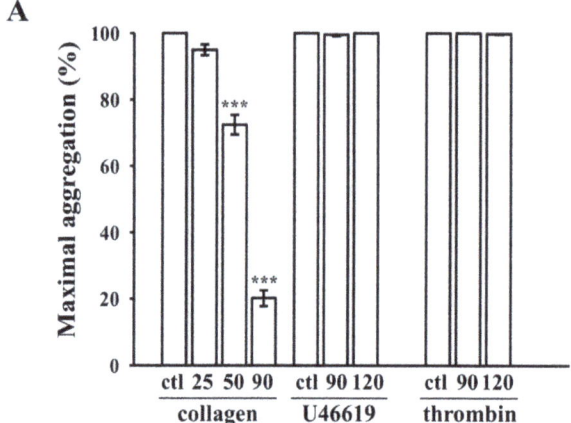

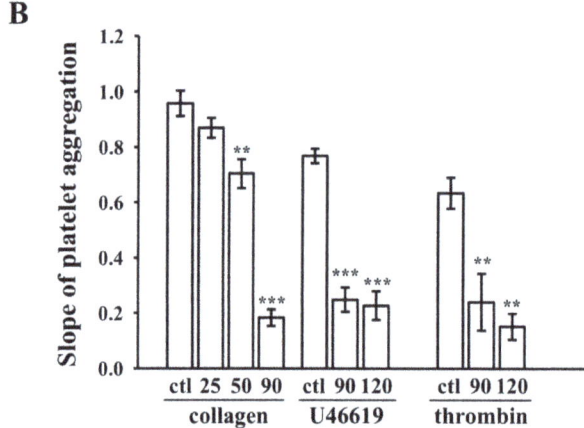

Figure 4. Maximal aggregation and slope of aggregation curves. (**A**) Concentration-response bar diagrams of the synthesized CDOTs, demonstrating their inhibitory activity for maximal aggregation (%). (**B**) Slope of platelet aggregation, as calculated from the linear part of the aggregation trace. Values represent mean ± SEM ($n = 4$). ** $p < 0.01$ and *** $p < 0.001$, compared with the control (ctl; PBS-treated) group.

3.4. Inhibition of PKC Activation (p47; Pleckstrin) and Akt, JNK1/2, and p38 MAPK Phosphorylation

We additionally investigated the mechanisms by which the CDOTs inhibited platelet aggregation. Their effects on PKC activation (p-p47) and Akt (protein kinase B), JNK1/2, and p38 MAPK phosphorylation are shown in Figure 5. The CDOTs (65 and 90 μM) significantly and, in a concentration-dependent manner, suppressed PKC activation in collagen-activated platelets, as shown in Figure 5A. Akt is a serine/threonine-specific protein kinase, which acts a major element in numerous cellular events, such as platelet activation, cell proliferation, apoptosis, and cell migration [26]. The CDOTs markedly inhibited collagen-induced Akt phosphorylation, as shown in Figure 5B. Furthermore, the CDOTs inhibited JNK1/2 and p38 MAPK [27] phosphorylation, which were elevated in collagen-stimulated platelets, as shown in Figure 5C,D, respectively.

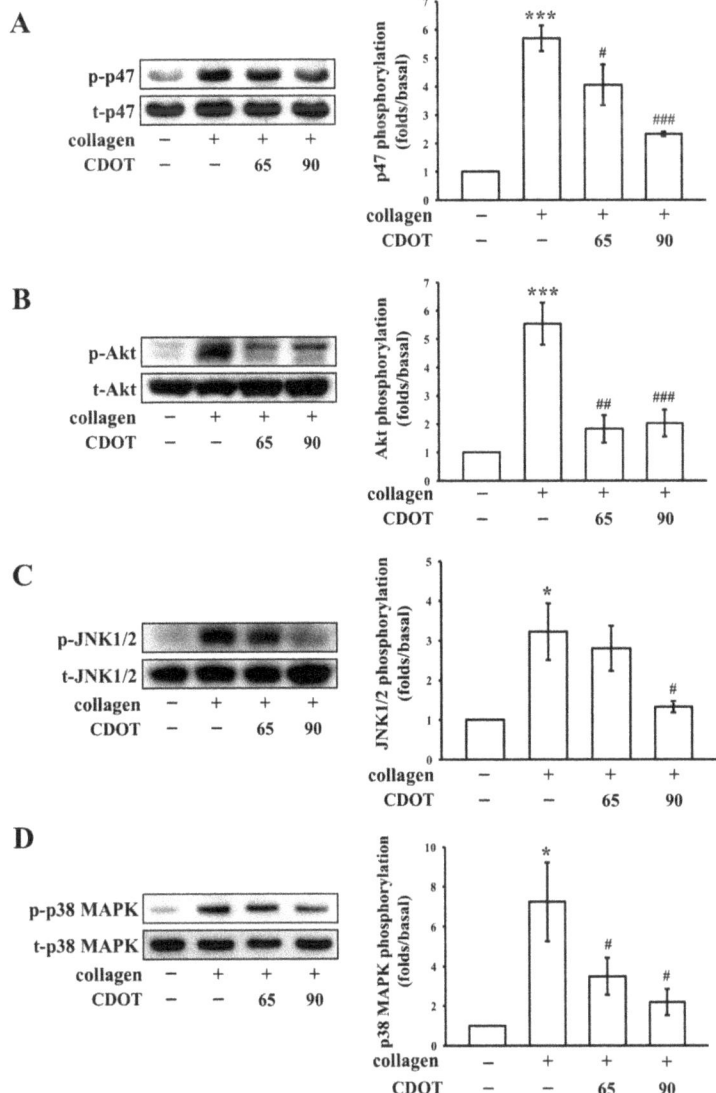

Figure 5. Effects of the CDOTs on PKC activation, and Akt, JNK1/2, and p38 MAPK phosphorylation in collagen-activated platelets. Washed platelets (1.2×10^9 cells/mL) were preincubated with PBS (control) or the synthesized CDOTs (65 and 90 μM), and subsequently, collagen (1 μg/mL) was added to trigger (**A**) PKC activation (p-p47) and (**B**) Akt, (**C**) JNK1/2, and (**D**) p38 MAPK phosphorylation. Values represent mean ± SEM ($n = 4$). * $p < 0.05$ and *** $p < 0.001$, compared with the control (PBS-treated) group. # $p < 0.05$, ## $p < 0.01$, and ### $p < 0.001$, compared with the collagen-treated group.

3.5. Effects of the CDOTs on Tail Bleeding and Mortality in Mice with ADP-Induced Pulmonary Thromboembolism

Bleeding is a common side-effect of the antiplatelet drugs used in this study. We evaluated the effects of the CDOTs on bleeding time via a tail transection model. The bleeding time was 65.3 ± 4.2 s ($n = 8$) in the control group, as shown in Figure 6A. After 10 min of intravenous administration of the

CDOTs (1 mg/kg), the bleeding time was 69.4 ± 5.7 s (n = 8). As is evident, the bleeding time was not significantly affected.

Further, we investigated mortality in mice with ADP-induced acute pulmonary thromboembolism treated with the CDOTs. The mortality rate of the animals with ADP-induced acute pulmonary thromboembolism (0.7 mg/g ADP) was 75% (i.e., deaths of 6 mice, n = 8); however, pretreatment with the CDOTs (1 mg/kg) considerably reduced the mortality rate to only around 25% (i.e., deaths of 2 mice, n = 8), as shown in Figure 6B. Hematoxylin–eosin was used to stain the lung tissues of the mice. As shown in Figure 6C, the control group exhibited severe pulmonary thrombosis (arrows), whereas the CDOTs (1 mg/kg) exerted substantial protective effects. Overall, these results showed that the synthesized CDOTs had an eminent antiplatelet effect in vivo without the side-effect of bleeding.

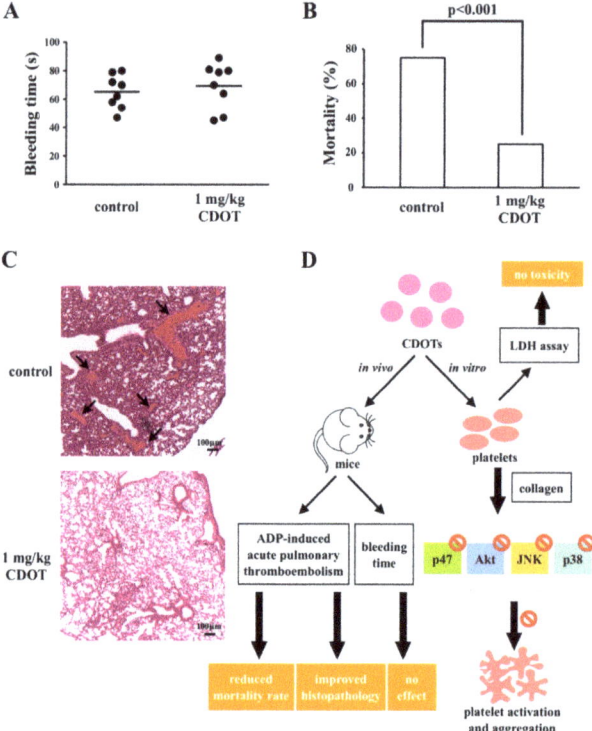

Figure 6. Effects of CDOTs on tail bleeding time and pulmonary thrombosis in experimental mice. (**A**) Bleeding time was measured through tail transection after 10 min of intravenous administration of PBS (control) or 1 mg/kg CDOTs. The bleeding time was continuously recorded until no sign of bleeding was observed for at least 10 s. (**B**) For the study of acute pulmonary thrombosis, PBS (control) or 1 mg/kg CDOTs was intravenously administered to mice, and ADP (0.7 mg/g) was then injected through the tail veins. (**C**) Pulmonary thrombosis (arrows) was observed by staining lung tissue sections with hematoxylin–eosin. Scale bar: 100 μm. Values represent mean ± SEM (n = 8). (**D**) Schematic illustration showing the inhibitory effect of CDOTs in human platelets. CDOTs potently inhibit human platelet activation by suppressing PKC activation and Akt, JNK1/2, and p38 MAPK phosphorylation without inducing cytotoxicity. CDOTs reduced the mortality in ADP-induced thromboembolic mice and did not affect bleeding tendency.

4. Discussion

CDOTs have extensively been applied in different fields for drug delivery. They have also been used in bioimaging and as effective biosensors for protein detection [14], considering their excellent biocompatibility, good water solubility, low toxicity, high photoluminescence, and high photostability. In this study, we synthesized a new type of CDOT from garlic. These nanoparticles were potent at hindering collagen-induced platelet aggregation and only reduced the slope of the aggregation curve (lag time) by U46619 and thrombin. Different physiological agonists (e.g., collagen, thrombin, and ADP) activated platelets. The primary activation of agonists may be enriched by the secondary activation induced by thromboxane A_2 from arachidonic acid or by ADP from the granules in platelets. In the case of blood vessel injury, platelets adhere to the subendothelial matrix (collagen), causing granule secretion and platelet activation. Collagen, a matrix protein which exists in the vascular subendothelium and vessel wall, acts as substrate for platelet adhesion and potent platelet stimulator. In this manner, collagen exerts as a key player in platelet activation.

To exclude the possible cytotoxic effects of the synthesized CDOTs on human platelets, we estimated the leakage of cytosolic LDH. LDH, a soluble cytoplasmic enzyme which occurs in nearly all cells is released into the extracellular space when the plasma membrane is injured. We found that the alteration between the control and platelets subjected with 200 μM CDOTs was not substantial, suggesting potential hemocompatibility. This result is concordant with that reported by Shrivastava et al. [28], who established that silver nanoparticles did not disturb platelet membrane integrity, even at concentrations as high as 500 μM. In addition, LDH release was not noticed from platelets after exposure to 0.9–3.5 nM silver nanoparticles [29]. Moreover, in a recent study, Hajtuch et al. [30] reported that functionalized silver nanoparticles, such as AgNPs-GSH, AgNPs-PEG, and AgNPs-LA, ranging in size from 2 to 3.7 nm, inhibited platelet aggregation without releasing LDH. The results pertaining to the effects of nanoparticles on platelets are inconsistent. Studies have found that gold nanoparticles are inert [31] or activate [32] platelets. Silver nanoparticles have been reported to induce platelet aggregation both in human platelets and in an animal models [33]. Huang et al. [34] demonstrated that silver nanoparticles coated with polyvinyl pyrrolidone and citrate had no significant effects on human platelet aggregation. In this study, we found that the synthesized CDOTs effectively inhibited collagen-triggered platelet aggregation. Consistent with our findings, Ragaseema et al. [35] reported the inhibitory effects of silver nanoparticles on platelet aggregation. In addition, Shrivastava et al. [28] found that silver nanoparticles condensed ADP- and collagen-induced platelet activation with a reduction in the slope of aggregation. These inconsistencies could be attributed to differences in size, stabilization, and functionalization, as well as the method of nanoparticle synthesis.

In the present study, the CDOTs evidently inhibited collagen-stimulated platelet activation, implying that they were effective in inhibiting platelet activation via a prominent phospholipase C (PLC)-dependent mechanism. PLC, belonging to a family of kinases, hydrolyzes phosphatidylinositol 4,5-bisphosphate to yield two chief secondary messengers: diacylglycerol and inositol trisphosphate. Diacylglycerol activates PKC-inducing pleckstrin (p47) phosphorylation and ATP release in activated platelets, whereas inositol trisphosphate elevates calcium influx [36]. The observed antiplatelet effects of the CDOTs could be a result of the inhibition of the PLC–PKC cascade, leading to the suppression of Akt and MAPK activation. Akt (a downstream regulator of PI3K)-knockout mice have been found to demonstrate impaired platelet activation [22]. Hence, Akt inhibition may be considered as striking antithrombotic targets. Conservative MAPKs are classified into ERK1/2, p38 MAPK, JNK1/2, and big MAPK (ERK5) [37]. ERK1/2, JNK1/2, and p38 MAPK participate in platelet activation [37]. MAPK presents in platelets linked to the mechanistic role of several antiplatelet agents [38]. Adam et al. [39] reported that JNK1 knockdown reduced platelet aggregation, with JNK1$^{-/-}$ platelets displaying abnormal platelet granule secretion, and led to defective thrombus formation in mice. p38 MAPK is associated with thrombus formation, as evidenced in p38 MAPK$^{-/-}$ mice [39,40]. Therefore, PKC, Akt, JNK1/2, and p38 MAPK are regarded as major targets of antiplatelet agents. A study demonstrated that silica nanoparticles induced expressions of the phosphorylated JNK and p38 MAPK, and suppressed ERK phosphorylation in human umbilical vein endothelial cells [41]. In the current study, the synthesized CDOTs markedly inhibited collagen-induced PKC, Akt, JNK1/2, and p38 MAPK phosphorylation in a concentration-dependent manner.

The GPVI receptor induces strong signaling through the protein tyrosine kinase pathways that results in the activation of PI3 and PLCγ and Ca^{2+} release. Since, in this study, CDOT effectively inhibited collagen-induced platelet aggregation, GPVI receptor-mediated inhibitory signaling pathways maybe involved in this anti-aggregatory effect. Thus, we believe that the inhibition of these signaling molecules by the CDOTs may lead to inhibitory effects on platelet activation. Miller et al. proved the hypothesis that the biological activity of nanoparticles may be dictated by their composition, size, and charge [42]. They found that human- or bovine-derived nanoparticles inhibited platelet aggregation induced by two different agonists—one that activates the thrombin receptor and the other that activates the collagen receptor—and they suggested that the inhibitory effects may be nonspecific, possibly by reducing platelet–platelet interactions or by binding to these or other surface receptors. Consistent with these discoveries, the current in vitro observation of the potent inhibitory effect of CDOTs in collagen-induced human platelet aggregation may be due to its inhibition of platelet–platelet interactions or by preventing binding with the collagen receptor. However, the detailed mechanisms of these hypotheses remain to be explored.

The intravenous administration of nanoparticles has been previously reported to substantially inhibit platelet aggregation in mice, indicative of their in vivo antiplatelet effects [28]. Furthermore, Shrivastava et al. [28] conducted tail-bleeding assays to determine the presence of any opposing effect on bleeding time and found that mice continued to live normally after nanoparticle administration. Similarly, Kim et al. [43] found that silver nanoparticles were nontoxic to rodents, and in another more relevant study, gold nanoparticles were observed to inhibit both thrombosis and considerably improve the survival rates of mice, without increasing the bleeding risk [44]. These results are consistent with those of this study, where even we found that CDOT administration reduced mortality in thromboembolic mice without prolonging the bleeding tendency.

5. Conclusions

CDOTs have recently gained much attention worldwide. Herein we found that the synthesized CDOTs could actively inhibit human platelet activation by suppressing PKC activation and Akt, JNK1/2, and p38 MAPK phosphorylation. Furthermore, there was no cytotoxicity in vitro. The in vivo study revealed that the CDOTs had an antithrombotic effect on the ADP-induced pulmonary thromboembolic mice model. CDOTs attenuate ADP-induced severe pulmonary thrombosis via the potential recovering of lung histopathology, reducing mortality and maintaining the normal bleeding tendency in mice. Altogether, our results suggest that a direct application of CDOTs may contribute to the development of new antiplatelet drugs for the treatment of arterial thromboembolic diseases.

Author Contributions: Conceptualization and designing of the projects, P.V. and J.-R.S.; performing experiments, T.-Y.L., P.T., K.-C.L., and H.-M.C.; original manuscript preparation, T.J. and J.-R.S.; nanoparticle synthesis, P.V.; writing—review and editing, all authors. All authors have read and agree to the published version of the manuscript.

Funding: This work was supported by grants from the Ministry of Science and Technology of Taiwan (MOST-106-2113-M-001-032, NSC 102-2113-M-002-009-MY3, MOST 107-2320-B-038-035-MY2, and MOST 108-2320-B-038-031-MY3) and Taipei Medical University (DP2-109-21121-01-N-08-03).

Conflicts of Interest: The authors declare no conflict of interest.

References

1. Watson, R.D.; Chin, B.S.; Lip, G.Y. Antithrombotic therapy in acute coronary syndromes. *Br. Med. J.* **2002**, *325*, 1348–1351. [CrossRef] [PubMed]
2. Baruah, D.B.; Dash, R.N.; Chaudhari, M.R.; Kadam, S.S. Plasminogen activators: A comparison. *Vasc. Pharmacol.* **2006**, *44*, 1–9. [CrossRef] [PubMed]

3. Almoosa, K. Is Thrombolytic therapy effective for pulmonary embolism. *Am. Fam. Physician* **2002**, *65*, 1097–1102. [PubMed]
4. Maxwell, M.J.; Yuan, Y.; Anderson, K.E.; Hibbs, M.L.; Salem, H.H.; Jackson, S.P. SHIP1 and Lyn Kinase negatively regulate integrin αIIbβ3 signaling in platelets. *J. Biol. Chem.* **2004**, *279*, 32196–32204. [CrossRef] [PubMed]
5. Scuri, M.; Chen, B.T.; Castranova, V.; Reynolds, J.S.; Johnson, V.J.; Samsell, L.; Walton, C.; Piedmonte, G. Effects of titanium dioxide nanoparticle exposure on neuroimmuneresponses in rat airways. *J. Toxicol. Environ. Health A* **2010**, *73*, 1353–1369. [CrossRef] [PubMed]
6. Buffle, J. The key role of environmental colloids/nanoparticles for the sustainability of life. *Environ. Chem.* **2006**, *3*, 155–158. [CrossRef]
7. Hasan, A.; Morshed, M.; Memic, A.; Hassan, S.; Webster, T.J.; Marei, H.E. Nanoparticles in tissue engineering: Applications, challenges and prospects. *Int. J. Nanomed.* **2018**, *13*, 5637–5655. [CrossRef]
8. Berry, J.P.; Arnoux, B.; Stanislas, G.; Galle, P.; Chretien, J. A microanalytic study of particles transport across the alveoli: Role of blood platelets. *Biomedicine* **1977**, *27*, 354–357.
9. Gaffney, A.M.; Santos-Martinez, M.J.; Satti, A.; Major, T.C.; Wynne, K.J.; Gun'ko, Y.K.; Annich, G.M.; Elia, G.; Radomski, M.W. Blood biocompatibility of surface-bound multi-walled carbon nanotubes. *Nanomedicine* **2015**, *11*, 39–46. [CrossRef]
10. Anselmo, A.C.; Modery-Pawlowski, C.L.; Menegatti, S.; Kumar, S.; Vogus, D.R.; Tian, L.L.; Chen, M.; Squires, T.M.; Sen Gupta, A.; Mitragotri, S. Platelet-like nanoparticles: Mimicking shape, flexibility, and surface biology of platelets to target vascular injuries. *ACS Nano* **2014**, *8*, 11243–11253. [CrossRef]
11. Roy, S.C.; Paulose, M.; Grimes, C.A. The effect of TiO2 nanotubes in the enhancement of blood clotting for the control of hemorrhage. *Biomaterials* **2007**, *28*, 4667–4672. [CrossRef]
12. Gayen, B.; Palchoudhury, S.; Chowdhury, J. Carbon dots: A mystic star in the world of nanoscience. *J. Nanomater.* **2019**, *2019*, 19. [CrossRef]
13. Lim, S.Y.; Shen, W.; Gao, Z. Carbon quantum dots and their applications. *Chem. Soc. Rev.* **2015**, *44*, 362–381. [CrossRef] [PubMed]
14. Zuo, J.; Jiang, T.; Zhao, X.; Xiong, X.; Xiao, S.; Zhu, Z. Preparation and application of fluorescent carbon dots. *J. Nanomater.* **2015**, *2015*, 1–13. [CrossRef]
15. Wang, R.; Lu, K.-Q.; Tang, Z.-R.; Xu, Y.-J. Recent progress in carbon quantum dots: Synthesis, properties and applications in photocatalysis. *J. Mater. Chem. A* **2017**, *5*, 3717–3734. [CrossRef]
16. Yan, X.; Zhao, Y.; Luo, J.; Xiong, W.; Liu, X.; Cheng, J.; Wang, Y.; Zhang, M.; Qu, H. Hemostatic bioactivity of novel Pollen Typhae Carbonisata-derived carbon quantum dots. *J. Nanobiotechnol.* **2017**, *15*, 60. [CrossRef] [PubMed]
17. Zhao, Y.; Zhang, Y.; Liu, X.; Kong, H.; Wang, Y.; Qin, G.; Cao, P.; Song, X.; Yan, X.; Wang, Q.; et al. Novel carbon quantum dots from egg yolk oil and their haemostatic effects. *Sci. Rep.* **2017**, *7*, 4452. [CrossRef]
18. Liu, X.; Wang, Y.; Yan, X.; Zhang, M.; Zhang, Y.; Cheng, J.; Lu, F.; Qu, H.; Wang, Q.; Zhao, Y. Novel phellodendri cortex (huang bo)-derived carbon dots and their hemostatic effect. *Nanomedicine* **2018**, *13*, 391–405. [CrossRef]
19. Fedel, M. Hemocompatibility of carbon nanostructures. *C J. Carbon Res.* **2020**, *6*, 12. [CrossRef]
20. Sheu, J.R.; Lee, C.R.; Lin, C.H.; Hsiao, G.; Ko, W.C.; Chen, Y.C.; Yen, M.H. Mechanisms involved in the antiplatelet activity of Staphylococcus aureus lipoteichoic acid in human platelets. *Thromb. Haemost.* **2000**, *83*, 777–784.
21. Lu, W.J.; Lee, J.J.; Chou, D.S.; Jayakumar, T.; Fong, T.H.; Hsiao, G.; Sheu, J.R. A novel role of andrographolide, an NF-κB inhibitor, on inhibition of platelet activation: The pivotal mechanisms of endothelial nitric oxide synthase/cyclic GMP. *J. Mol. Med.* **2011**, *89*, 1261–1273. [CrossRef] [PubMed]
22. Zhu, C.; Zhai, J.; Dong, S. Bifunctional fluorescent carbon nanodots: Green synthesis via soy milk and application as metal-free electrocatalysts for oxygen reduction. *Chem. Commun. (Camb.)* **2012**, *48*, 9367–9369. [CrossRef]
23. Alkian, I.; Prasetio, A.; Anggara, L.; Karnaji; Fonisyah, M.H.; Rizka, Z.M.; Widiyandari, H. A facile microwave-assisted synthesis of carbon dots and their application as sensitizers in nanocrystalline TiO2 solar cells. *J. Phys. Conf. Ser.* **2019**, *1204*, 012093. [CrossRef]

24. Wu, Z.L.; Zhang, P.; Gao, M.X.; Liu, C.F.; Wang, W.; Leng, F.; Huang, C.Z. One-pot hydrothermal synthesis of highly luminescent nitrogen-doped amphoteric carbon dots for bioimaging from bombyx mori silk—Natural proteins. *J. Mater. Chem. B* **2013**, *1*, 2868–2873. [CrossRef] [PubMed]
25. Zhao, S.; Lan, M.; Zhu, X.; Xue, H.; Ng, T.W.; Meng, X.; Lee, C.S.; Wang, P.; Zhang, W. Green synthesis of bifunctional fluorescent carbon dots from garlic for cellular imaging and free radical scavenging. *ACS Appl. Mater. Interfaces* **2015**, *7*, 17054–17060. [CrossRef]
26. Woulfe, D.S. Akt signaling in platelet and thrombosis. *Expert Rev. Hematol.* **2010**, *3*, 81–91. [CrossRef]
27. Bugaud, F.; Nadal-Wollbold, F.; Levy-Toledano, S.; Rosa, J.P.; Bryckaert, M. Regulation of c-jun-NH2 terminal kinase and extracellular-signal regulated kinase in human platelets. *Blood* **1999**, *94*, 3800–3805. [CrossRef] [PubMed]
28. Shrivastava, S.; Bera, T.; Singh, S.K.; Singh, G.; Ramachandrarao, P.; Dash, D. Characterization of antiplatelet properties of silver nanoparticles. *ACS Nano* **2009**, *3*, 1357–1364. [CrossRef]
29. Krishnaraj, R.N.; Berchmans, S. In vitro antiplatelet activity of silver nanoparticles synthesized using the microorganism Gluconobacter roseus: An AFM-based study. *RSC Adv.* **2013**, *3*, 8953–8959. [CrossRef]
30. Hajtuch, J.; Hante, N.; Tomczyk, E.; Wojcik, M.; Radomski, M.W.; Santos-Martinez, M.J.; Inkielewicz-Stepniak, I. Effects of functionalized silver nanoparticles on aggregation of human blood platelets. *Int. J. Nanomed.* **2019**, *14*, 7399–7417. [CrossRef]
31. Love, S.A.; Thompson, J.W.; Haynes, C.L. Development of screening assays for nanoparticle toxicity assessment in human blood: Preliminary studies with charged Au nanoparticles. *Nanomedicine* **2012**, *7*, 1355–1364. [CrossRef] [PubMed]
32. Deb, S.; Patra, H.K.; Lahiri, P.; Dasgupta, A.K.; Chakrabarti, K.; Chaudhuri, U. Multistability in platelets and their response to gold nanoparticles. *Nanomedicine* **2011**, *7*, 376–384. [CrossRef] [PubMed]
33. Jun, E.A.; Lim, K.M.; Kim, K.; Bae, O.N.; Noh, J.Y.; Chung, K.H.; Chung, J.H. Silver nanoparticles enhance thrombus formation through increased platelet aggregation and procoagulant activity. *Nanotoxicology* **2011**, *5*, 157–167. [CrossRef] [PubMed]
34. Huang, H.; Lai, W.; Cui, M.; Liang, L.; Lin, Y.; Fang, Q.; Liu, Y.; Xie, L. An evaluation of blood compatibility of silver nanoparticles. *Sci. Rep.* **2016**, *6*, 255180. [CrossRef] [PubMed]
35. Ragaseema, V.M.; Unnikrishnan, S.; Kalliyana Krishnan, V.; Krishnan, L.K. The antithrombotic and antimicrobial properties of PEG-protected silver nanoparticle coated surfaces. *Biomaterials* **2012**, *33*, 3083–3092. [CrossRef] [PubMed]
36. Singer, W.D.; Brown, H.A.; Sternweis, P.C. Regulation of eukaryotic phosphatidylinositol-specific phospholipase C and phospholipase D. *Ann. Rev. Biochem.* **1997**, *66*, 475–509. [CrossRef]
37. Fan, X.; Wang, C.; Shi, P.; Gao, W.; Gu, J.; Geng, Y.; Yang, W.; Wu, N.; Wang, Y.; Xu, Y.; et al. Platelet MEKK3 regulates arterial thrombosis and myocardial infarct expansion in mice. *Blood Adv.* **2018**, *2*, 1439–1448. [CrossRef]
38. Mazharian, A.; Roger, S.; Berrou, E.; Adam, F.; Kauskot, A.; Nurden, P.; Jandrot-Perrus, M.; Bryckaert, M. Protease-activating receptor-4 induces full platelet spreading on a fibrinogen matrix: Involvement of ERK2 and p38 and Ca2+ mobilization. *J. Biol. Chem.* **2007**, *282*, 5478–5487. [CrossRef]
39. Adam, F.; Kauskot, A.; Nurden, P.; Sulpice, E.; Hoylaerts, M.F.; Davis, R.J.; Rosa, J.P.; Bryckaert, M. Platelet JNK1 is involved in secretion and thrombus formation. *Blood* **2010**, *115*, 4083–4092. [CrossRef]
40. Adam, F.; Kauskot, A.; Rosa, J.P.; Bryckaert, M. Mitogen-activated protein kinases in hemostasis and thrombosis. *J. Thromb. Haemost.* **2008**, *6*, 2007–2016. [CrossRef]
41. Guo, C.; Xia, Y.; Niu, P.; Jiang, L.; Duan, J.; Yu, Y.; Zhou, X.; Li, Y.; Sun, Z. Silica nanoparticles induce oxidative stress, inflammation, and endothelial dysfunction in vitro via activation of the mapk/nrf2 pathway and nuclear factor-kappab signaling. *Int. J. Nanomed.* **2015**, *10*, 1463–1477. [CrossRef] [PubMed]
42. Miller, V.M.; Hunter, L.W.; Chu, K.; Kaul, V.; Squillace, P.D.; Lieske, J.C.; Jayachandran, M. Biologic nanoparticles and platelet reactivity. *Nanomedicine* **2009**, *4*, 725–733. [CrossRef] [PubMed]
43. Kim, Y.S.; Kim, J.S.; Cho, H.S.; Rha, D.S.; Kim, J.M.; Park, J.D.; Choi, B.S.; Lim, R.; Chang, H.K.; Chung, Y.H.; et al. Twenty-eight-day oral toxicity, genotoxicity, and gender-related tissue distribution of silver nanoparticles in Sprague-Dawley rats. *Inhal. Toxicol.* **2008**, *20*, 575–583. [CrossRef] [PubMed]
44. Tian, Y.; Zhao, Y.; Zheng, W.; Zhang, W.; Jiang, X. Antithrombotic functions of small molecule-capped gold nanoparticles. *Nanoscale* **2014**, *6*, 8543–8550. [CrossRef]

 © 2020 by the authors. Licensee MDPI, Basel, Switzerland. This article is an open access article distributed under the terms and conditions of the Creative Commons Attribution (CC BY) license (http://creativecommons.org/licenses/by/4.0/).

Article

Covalent Decoration of Cortical Membranes with Graphene Oxide as a Substrate for Dental Pulp Stem Cells

Roberta Di Carlo [1,†], Susi Zara [1,†], Alessia Ventrella [1,†], Gabriella Siani [1], Tatiana Da Ros [2], Giovanna Iezzi [3], Amelia Cataldi [1] and Antonella Fontana [1,*]

1 Department of Pharmacy, University "G. d'Annunzio", Via dei Vestini, 66100 Chieti, Italy; roberta.dicarlo@unich.it (R.D.C.); susi.zara@unich.it (S.Z.); alessia.ventrella@unich.it (A.V.); gabriella.siani@unich.it (G.S.); amelia.cataldi@unich.it (A.C.)
2 Department of Chemical and Pharmaceutical Sciences, University of Trieste, Piazzale Europa 1, 34127 Trieste, Italy; daros@units.it
3 Department of Medical, Oral and Biotechnological Sciences, University "G. d'Annunzio", Via dei Vestini, 66100 Chieti, Italy; giovanna.iezzi@unich.it
* Correspondence: fontana@unich.it; Tel.: +39-0871-3554790
† These authors contributed equally to this work.

Received: 29 March 2019; Accepted: 8 April 2019; Published: 12 April 2019

Abstract: (1) Background: The aim of this study was to optimize, through a cheap and facile protocol, the covalent functionalization of graphene oxide (GO)-decorated cortical membrane (Lamina®) in order to promote the adhesion, the growth and the osteogenic differentiation of DPSCs (Dental Pulp Stem Cells); (2) Methods: GO-coated Laminas were fully characterized by Scannsion Electron Microscopy (SEM) and Atomic Force Microscopy (AFM) analyses. In vitro analyses of viability, membrane integrity and calcium phosphate deposition were performed; (3) Results: The GO-decorated Laminas demonstrated an increase in the roughness of Laminas, a reduction in toxicity and did not affect membrane integrity of DPSCs; and (4) Conclusions: The GO covalent functionalization of Laminas was effective and relatively easy to obtain. The homogeneous GO coating obtained favored the proliferation rate of DPSCs and the deposition of calcium phosphate.

Keywords: graphene oxide; covalent functionalization; cortical membranes; calcium phosphate deposition

1. Introduction

In this study, we focused our interest on cortical membranes, commonly used in oral surgical procedures, in order to improve their features thanks to a covalent enrichment with graphene oxide (GO). In particular, we used a type of cortical membrane, namely Osteobiol® Lamina provided by Tecnoss. Laminas, created by a registered trademark of Tecnoss, are made up of cortical bone of heterologous origin and are demonstrated to increase the rate of physiological resorption of the material [1]. Laminas have the compactness of bone tissue as well as a flexibility and adaptability that derives from the superficial decalcification process tuned for their preparation [1]. These materials are therefore used to improve bone tissue regeneration in cases in which it is important to reserve a space [2]. It is important to note that these tissue-derived materials are generally brittle and characterized by a low resistance to fracture. These drawbacks were overcome by enriching the original material, i.e., hydroxyapatite, with different additives such as alginate/chitosan [3], titania [4] or carbon nanotubes (CNT) [5].

Nowadays, graphene has emerged as a great alternative material for applications in biomedical and regenerative engineering. Graphene is a two-dimensional (2D) carbon-based material which has

sp^2 bonded carbon atoms arranged in a honeycomb lattice structure, with extraordinary electrical, physical, and optical properties [6]. Since its discovery, graphene and its derivatives have been widely investigated for the development of electrical devices and for biomedical applications such as drug delivery systems, biosensors, and regenerative therapies [7]. Mechanically, graphene, despite its flexibility, appears to be one of the strongest materials ever tested. [6] It is transparent, able to conduct electricity and heat better than metals [8], chemically inert, and stable [9]. An increasing number of studies have recently focused on the expansion of new potential applications of graphene nanomaterials, with the aim to highlight the benefits of their use and to improve the application of these nanomaterials [10].

Despite these properties, graphene has a very low solubility in both organic and aqueous solvents. For this reason, hydrophilic graphene derivatives, namely Graphene Oxide (GO), have been widely used and tested for pharmaceutical and biomedical applications. GO is hydrophilic, does not tend to form aggregates, and is highly and homogeneously dispersible in water. GO has been demonstrated to be a biocompatible material whose limited cytotoxicity depends on final concentration, shape, sheet size, dispersibility, and degree of surface functionalization [10]. GO has been investigated for its ability to enhance the proliferation and differentiation of several types of stem cells [11].

The aim of this study was to achieve the covalent functionalization of Laminas, by exploiting, via a simple, cheap, and effective protocol, the capacity of oxygenated groups of GO to interact with cortical membrane surfaces. Indeed, previously investigated GO-coatings [12–14] were obtained by simply depositing GO on the elected substrates and therefore exploiting weak London, Van der Waals, or hydrogen-bonding interactions. The concentrations of GO chosen are those that, in preliminary biological assays and in previous studies [13,14], demonstrated not to be toxic for fibroblast cells and favor osteogenic differentiation in dental pulp stem cells (DPSCs) on collagen membranes. The idea is to demonstrate the ability of graphene oxide to improve Laminas biological properties as well as promote the adhesion, the growth and the osteogenic differentiation of DPSCs (Dental Pulp Stem Cells). DPSCs were chosen because of the easy access to the site collection. Besides, DPSCs have an extensive differentiation ability and their capacity to interact with biomaterials makes them ideal for tissue reconstruction [11].

2. Materials and Methods

2.1. Materials

Cortical membranes (0.5 × 0.5 × 0.2 cm) (Ostebiol® Lamina, Tecnoss) were a gift of Tecnoss dental s.r.l. Pianezza (TO), Italy. GO was purchased from Graphenea, San Sebastian, Spain as an aqueous solution of 4 mg/mL GO. This solution was diluted at the elected concentration and bath ultrasonicated for 10 min (Elmasonic P60H, 37 kHz, 180 W) before use.

All other reagents were product of analytical grade from Merck KGaA, Darmstadt, Germany and they were used as received.

2.2. Enrichment with Graphene Oxide

In order to prepare GO-coated Laminas, a protocol of covalent functionalization was optimized. Firstly, the cortical membranes was activated by using a UV/ozone lamp (PSD-UV4 Novascan UV Ozone System Base model, Novascan Technologies, Boone, NC, USA) for 15 min on each side. This permits the subsequent coating with the functional groups. Secondly, the Laminas were dipped in 1 M ethanolic solution of 3-aminopropyl triethoxysilane (APTES, commercial sample from Merck KGaA, Darmstadt, Germany) for 3 h to obtain a thin, stable aminosilane layer on the activated membranes. The so obtained aminosilane-functionalized membranes were rinsed with ethanol and deionized water. Thirdly, these aminosilane-functionalized cortical membranes were dipped in graphene oxide aqueous solution of two different concentrations, 5 or 10 µg/mL. In particular, 4 mL of homogenous dispersion of GO in water were added to 10 cortical membranes (ca. 21 mg) in a baker. The GO solution was left into contact with samples overnight. Finally, membranes were left to dry at room temperature.

Samples were transferred in a 48 multi-well plate for the in vitro tests.

2.3. Sterilization of Cortical Membranes

Both pure and GO-coated Laminas were irradiated by using UV irradiation in the Herasafe KS 15, class II, type A2 biological safe cabinet (Thermo Fisher Scientific, North Logan, UT, USA) for 1 h on each side in order to sterilize the specimens.

2.4. Apparatus for Chemico-Physical Characterization of Laminas

Thermo-gravimetric analyses (TGA) were recorded on a TGA Q500 (TA Instruments, New Castle, DE, USA) on ca. 12 mg sample. The runs were performed under nitrogen atmosphere by equilibrating the samples at 100 °C for 20 min, following a ramp at 10 °C/min up to 800 °C.

The morphology of Lamina and GO-coated Laminas was evaluated by Atomic Force Microscopy (AFM), using a Multimode 8 Bruker AFM microscope (Bruker, Milan, Italy) coupled with a Nanoscope V controller and commercial silicon tips (RTESPA 300, resonance frequency of 300 kHz and nominal elastic constant of 40 N·m^{-1}) with a typical apex radius of 8 nm in Peak Force and ScanAsyst™ in air mode.

By using this mode, it was possible, from the height panel, to calculate roughness and, from the force curves recorded at various points, to calculate the Young's modulus. In particular, NanoScope Analysis software 1.8 enables to select the force curves registered at each point of the scanned surface and, from each force curve, to calculate the Young's modulus by fitting the linear part of the retracting curve via a hertzian model. The deflection sensitivity and tip radius were calibrated, prior to use, against standard sapphire.

2.5. Isolation and Culture of DPSCs

The Local Ethical Committee of the University "G. d'Annunzio" Chieti-Pescara approved the project (approval number 1173, date of approval 31/03/2016), in agreement with the Declaration of Helsinki. Dental pulps were extracted from third molars derived from young male and female people (age range 18–28 years) which underwent surgical procedures. All patients signed informed consent. The study involved only impacted teeth without dental pathologies. After the extraction, the surrounding tissues were mechanically eliminated and processed as reported in our previous work [15].

Samples were rinsed with phosphate-buffered saline (PBS), maintained in Minimum Essential Medium (α-MEM) (Merck KGaA, Darmstadt, Germany) supplemented with 10% of Foetal Bovine Serum (FBS) and 1% antibiotics (penicillin/streptavidin mixture, EuroClone S.p.A, Milan, Italy) and sent to the laboratory for stem cells extraction [15]. When cells covered 80–90% of the flask area (subconfluence condition) they were subcultured. Antigen expression of CD29, CD45, CD105, CD73 CD90 and SSEA-4 was checked by flow cytometry [15].

2.6. DPSCs Culture on Laminas

Cells from the fourth passage (Figure S1 of the Supporting Information) were seeded on Laminas, 10,000 cells/cm^2 were used and cultured up to 28 days. Two hundred twenty Laminas for each experiment were used, fifty-five Laminas were used for each experimental point. Experiments were repeated for three times. At the established times cells were harvested and processed for the required analyses. The cells were cultured in α-MEM medium supplemented with differentiation factors such as 10 nM dexamethasone, 0.2 mM ascorbic acid and 10 mM β-glycerophosphate, as reported elsewhere [16,17].

2.7. Scannsion Electron Microscopy (SEM) Analysis

Samples were fixed with 1.25% glutaraldehyde in 0.1 M cacodylate buffer for 30 min, dehydrated through alcohol ascending series and then dried with hexamethyldisilazane followed by gold-coating. All micrographs were obtained at 15 kV on compact desktop Phenom XL SEM microscope.

2.8. Alamar Blue Cell Viability Assay

The Alamar blue test was performed in triplicate for each experimental sample at each experimental time. Cells viability was measured after 3, 7, 14 and 28 days of culture. The test is based on the reduction of Alamar blue reagent (Thermo Scientific, Rockford, IL, USA), performed only by viable cells, into a red product. At established experimental times the medium was replaced by a new one added with 10% of Alamar blue reagent, incubated for 4 h at 37 °C. A spectrophotometric reading at 570 and 600 nm wavelength was performed. The negative control was established as the value obtained without cells. The percentage reduction of Alamar blue reagent was calculated following the manufacturer instructions.

2.9. Lactate Dehydrogenase (LDH) Cytotoxicity Assay

To evaluate biomaterial cytotoxicity, LDH release into the medium was measured by means of CytoTox 96 non-radioactive cytotoxicity assay (Promega, Madison, WI, USA) at each time point (3, 7, 14 and 28 days). The LDH leakage in each well was normalized to the lysis value obtained in a lysis well of the same experimental point in which a lysis solution was added to the medium

2.10. Alizarin Red S (ARS) Staining

Alizarin red S is a calcium-sensing dye. DPSCs, differentiated towards the osteogenic phenotype, are able to deposit and to induce the mineralization of extracellular matrix rich in calcium phosphate, which can be identified by Alizarin red S. Calcium deposits are detectable for their bright orange-red color.

The DPSCs in each well were rinsed twice with PBS, PBS was discarded and DPSCs were fixed in paraformaldehyde 4% for 15 min at room temperature and then washed with deionized water. Alizarin red S staining solution 40 mM (Merck KGaA, Darmstadt, Germany) was added to each well and probed for 20 min at room temperature (RT) on a shaker. The wells were washed for five times in deionized water. Calcium deposits, stained in orange-red, were dissolved as follows: 10% acetic acid was added under shaking for 30 min. Laminas were scraped, the liquid containing deposits was collected and vortexed in a tube. Previously heated warm mineral oil (Merck KGaA, Darmstadt, Germany) was added, the tube maintained on ice for 5 min and eventually centrifuged at 20,000 g for 15 min. The supernatant was discarded and 10% ammonium hydroxide (Merck KGaA, Darmstadt, Germany) was added. The final solution was analyzed by a spectrophotometric reading performed at 405 nm wavelength.

2.11. Statistical Analysis

SPSS software version 16.0 (SPSS, Inc., Chicago, IL, USA) (Statistical Package for Social Science) and GraphPad Prism 5 were used to perform statistical analysis. Data were evaluated using one-way analysis of variance followed by the Tukey-Kramer post-hoc test. The results were expressed as the mean ± standard deviation (SD). $P < 0.05$ was considered to indicate a statistically significant difference.

3. Results

Laminas were enriched with GO at two different concentrations, 5 and 10 µg/mL. Photos of the obtained enriched cortical membranes are reported in Figure 1.

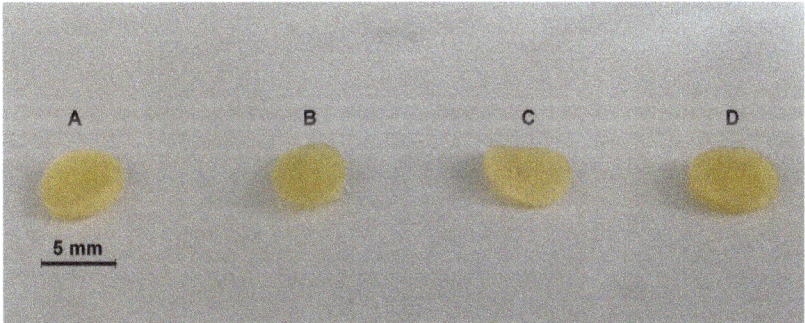

Figure 1. Photographs of (**A**) pure Lamina, (**B**) Lamina functionalized with 3-aminopropyl triethoxysilane (APTES) (see Experimental Section 2.2), (**C**) Lamina enriched with 5 µg/mL graphene oxide (GO) and (**D**) Lamina enriched with 10 µg/mL GO.

We tried to evaluate the amount of GO covalently attached to the functionalized Lamina by using TGA (Figure 2). While the GO sample presents the common behavior with an important weight loss at 200 °C, the Laminas profiles show a consistent weight loss at around 325 °C as for 5 µg/mL GO and 10 µg/mL GO (330 °C), even though for the last preparation a small implement of stability can be appreciated up to 270 °C with a difference in weight loss of 1.3% (9.1% vs. 10.4% in the case of control and 5 µg/mL GO).

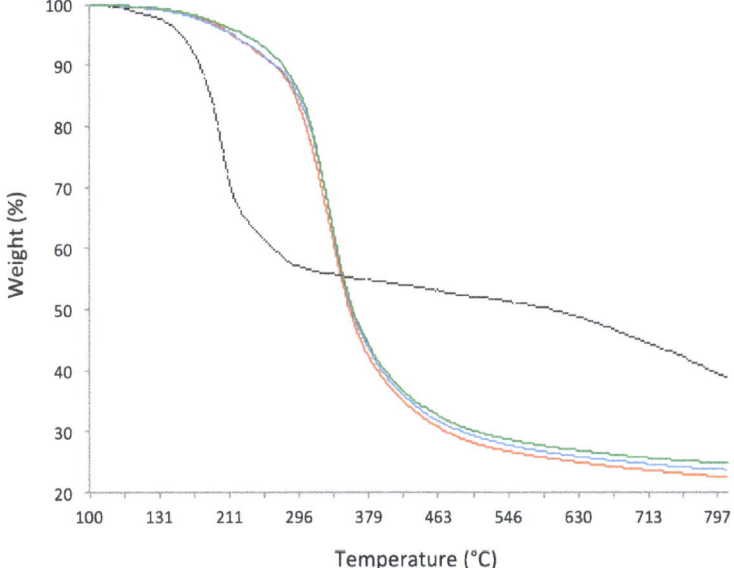

Figure 2. Thermo-gravimetric analyses (TGA) of graphene oxide (black curve), bare Lamina (red curve), 5 µg/mL GO enriched Lamina (blue curve) and Lamina enriched with 10 µg/mL GO (green curve).

Bare and GO-enriched cortical membranes were analyzed by using AFM. In Figure 3, topographical and tridimensional micrographs as well as peak force error images of bare and GO-enriched (5 µg/mL and 10 µg/mL) cortical membranes are reported. From the images reported the changes of the topography of the surface on enrichment with GO are evident.

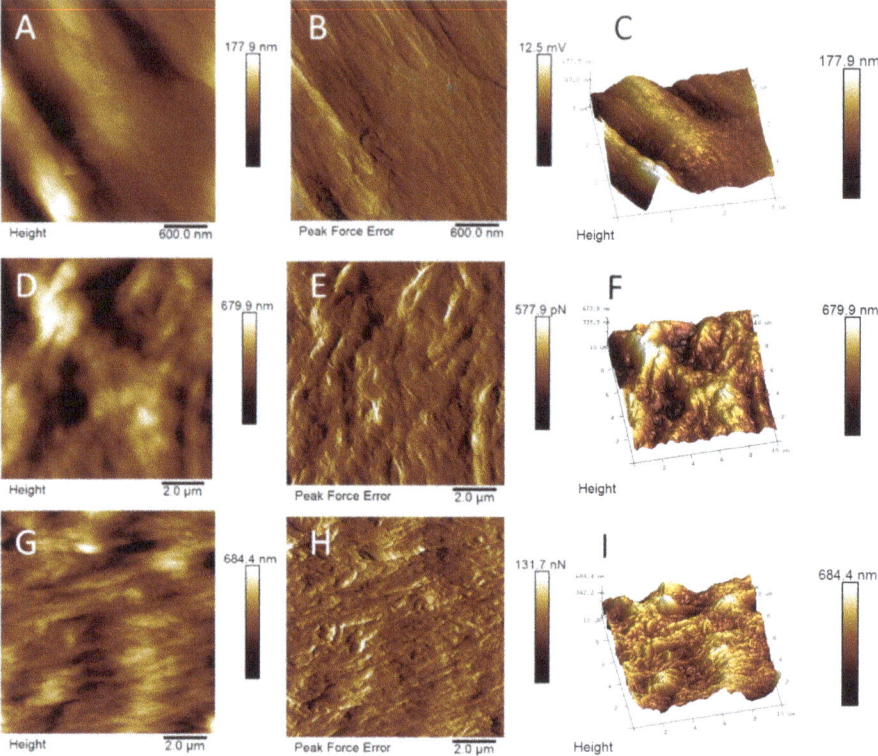

Figure 3. (**A,D,G**) Topographical, (**B,E,H**) Peak force error and (**C,F,I**) Three-dimensional Atomic Force Miscroscopy (AFM) images of bare Lamina (upper line), Lamina enriched with 5 µg/mL of GO (central line) and Lamina enriched with 10 µg/mL GO (bottom line).

From the height panel the Nanoscope analysis 1.8 software (Bruker, Milan, Itay) is able to recover the roughness indexes (i.e., the root-mean square roughness, Rq; the mean absolute value of the surface high deviations, Ra; i.e., the distance between the highest and lowest data points in the image, Rmax; the root-mean square of the surface slope, Sdq, and the ratio between the developed and the planar area, Sdr). We calculated these indexes for the bare membrane as the mean values of roughness recovered from two panels with a total surface area of 18 µm², were Rq = 53.0 ± 10.2 nm, Ra = 43.3 ± 9.3 nm, Rmax 255.5 ± 20.5 nm, Sdq 12.7 ± 1.6°, and Sdr 2.5 ± 0.5%. The roughness indexes, calculated as the mean values recovered from three panels with a total surface area of 300 µm², were Rq = 216.0 ± 21.2 nm, Ra = 175.0 ± 21.9 nm, Rmax 1303.3 ± 96.3 nm, Sdq 21.4 ± 1.8°, and Sdr 7.16 ± 1.0% for the GO-coated sample with 5 µg/mL and Rq = 254.7 ± 56.12 nm, Ra = 205.7 ± 46.5 nm, Rmax 1311.0 ± 282.0 nm, Sdq 25.9 ± 10.9°, and Sdr 11.5 ± 7.6% for the GO-coated sample with 10 µg/mL (See Supporting Information, Figures S8–S10).

Scansion electron microscopy (SEM) experiments (Figure 4) allowed to observe morphology differences in the investigated Laminas. As seen from SEM images (compare Figures 4B and 4A) the covalent functionalization with amino silane brought about a significant deformation of Lamina Surface. The subsequent coating with GO restored the typical layered structure of GO [13], with layered regions increasing on increasing GO concentration (compare Figures 4C and 4D).

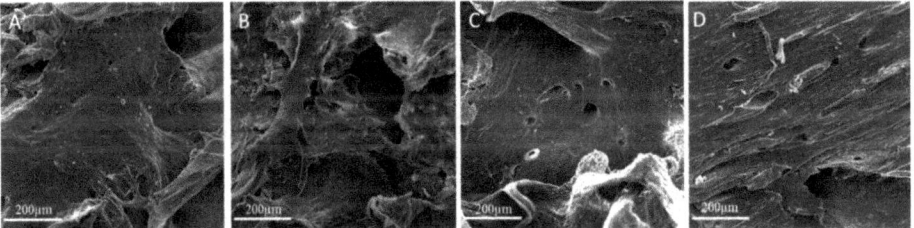

Figure 4. Scansion electron microscopy (SEM) images of (**A**) bare Lamina, (**B**) APTES-treated Lamina, (**C**) 5 µg/mL GO-coated Lamina and (**D**) 10 µg/mL GO-coated Lamina. Magnification 3000×. Scale bar: 200 µm.

DPSCs were cultured on Laminas with medium containing differentiation factors up to 28 days; 3, 7, 14, and 21 days were chosen as experimental times.

Before starting the evaluation of the biological parameters, an SEM analysis, after 7 and 14 days of culture, was performed in order to evaluate DPSCs morphology, spread and adhesion on Laminas. After 7 days of culture, cells are detectable on all the observed experimental points: DPSCs cultured on control Laminas are flat and spread throughout the surface, some granules of inorganic matrix are starting to be deposited. DPSCs cultured on APTES-treated Laminas appear isolated, with short cytoplasmic extensions, probably suffering and they do not cover all the surface of the Lamina. DPSCs grown on both 5 µg/mL GO- and 10 µg/mL GO-coated Laminas form a uniform cell layer on the biomaterial, they appear completely flat and in close contact with each other; an isolated cell is not recognizable. White granules of inorganic matrix can be identified especially on 5 µg/mL GO-coated Lamina. The same trend is revealed after 14 days of culture (Figure 5).

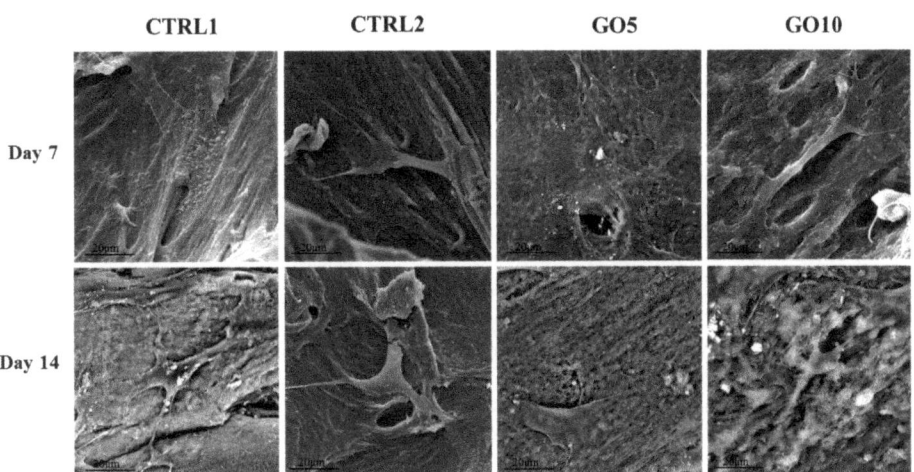

Figure 5. SEM images of Dental Pulp Stem Cells (DPSC)) cultured on bare Laminas (CTRL1), APTES-treated (CTRL2), 5 µg/mL GO-coated (GO5) and 10 µg/mL GO-coated (GO10) Laminas for 7 and 14 days. Magnification 3000×.

Cell viability was measured by means of Alamar Blue Assay after 3, 7, 14, and 28 days. After 3 days of culture the viability level does not show any significant difference among the tested experimental points, whereas after 7 days of culture an appreciable increase in viability level is recordable when DPSCs are cultured on GO-enriched Laminas. In particular, the cell viability is almost doubled for DPSCs cultured on 5 µg/mL GO-coated Laminas with respect to the control and it is comparably high for DPSCs cultured on 10 µg/mL GO-coated Laminas. Both the percentage of Alamar Blue reduction recorded on 5 µg/mL GO- and 10 µg/mL GO-coated membranes are statistically significant with respect to the control ($p < 0.001$). The metabolic activity of cell cultured on control and on GO-coated Laminas further augments (Figure 6) until 14 days of culture. By day 14, the number of viable cells reach a plateau, suggesting that those surfaces are advancing into confluence. On the other hand, it is worth noting that the proliferation rate of DPSCs cultured on APTES-treated Laminas is much lower, reaching the maximum percentage of Alamar Blue reduction at 28 days, when the difference with the other samples cancels out.

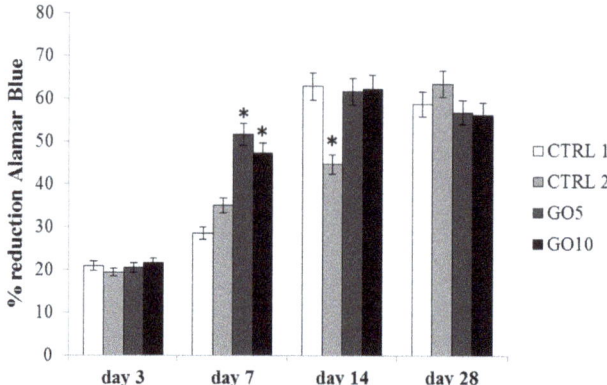

Figure 6. Alamar blue assay in DPSC cultured on bare Laminas (CTRL1), APTES-treated (CTRL2), 5 µg/mL GO-coated (GO5) and 10 µg/mL GO-coated (GO10) Laminas for 3, 7, 14, and 28 days. Forty Laminas were used for each experimental point, ten Laminas per experimental time. The histogram represents Alamar blue reduction percentage, data shown are the mean (±SD) of three separate experiments. Zero time % reduction Alamar Blue: 15.72%; * Day 7: GO5 and GO10 Laminas vs. control (CTRL1) Laminas $p < 0.001$; Day 14 control Laminas, GO5-coated and GO-10 coated Laminas vs. APTES-treated Laminas (CTRL2) $p < 0.001$.

The cytotoxicity of the biomaterial was evaluated through LDH assay by measuring the percentage of released LDH within the culture medium after 3, 7, 14, and 28 days of culture. After 3 days of culture the cytotoxicity level is higher than 70% for all tested samples except for DPSCs cultured on 10 µg/mL GO-coated Laminas. In fact, this sample shows a released LDH percentage significantly lower than that measured for the three others samples. After 7 days of culture the cytotoxicity level starts to decrease with respect to that measured after 3 days of culture in all the investigated samples except for DPSCs cultured on APTES-treated Laminas. In fact, the released LDH percentage for this sample appears still higher than 70%, whereas the percentage decreases under 60% for cells grown on control Laminas and under 40% for DPSCs cultured on GO-coated Laminas. The cytotoxicity level does not change thereafter for GO-coated samples and a statistically significant reduction of released LDH percentage is detected for DPSCs grown on 10 µg/mL GO-coated Laminas with respect to the control (Figure 7). Again DPSCs cultured on APTES-treated Laminas show the highest released LDH percentage (>50%) among the four Laminas investigated.

Bone matrix deposition was measured through Alizarin Red staining, a calcium-sensing dye able to identify extracellular quantities of calcium phosphate. Alizarin Red staining was performed after 21 and 28 days of culture. After 21 days of culture, a marked decrease of synthetized calcium phosphate could be detected in DPSCs cultured on APTES-treated Laminas compared with the control and with GO-coated Laminas. Conversely, after 28 days of culture, a statistically significant increase in calcium phosphate deposition is detected in DPSCs cultured on 5 µg/mL GO-enriched Laminas with respect to all other tested samples (Figure 8).

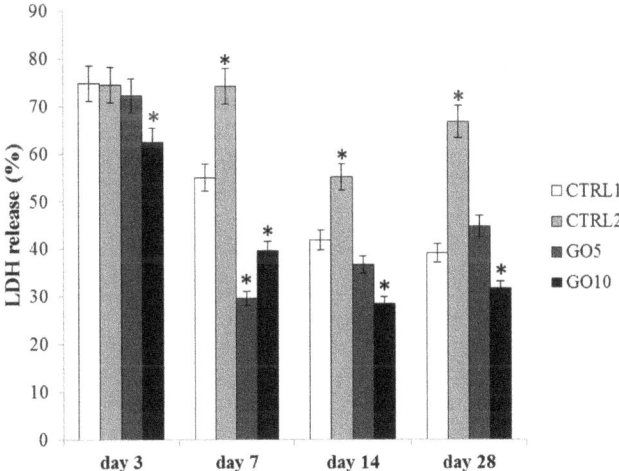

Figure 7. Lactate Dehydrogenase (LDH) assay of DPSC cultured on bare Laminas (CTRL1), APTES-treated (CTRL2), 5 µg/mL GO-coated (GO5) and 10 µg/mL GO-coated (GO10) Laminas for 3, 7, 14, and 28 days. Forty Laminas were used for each experimental point, ten Laminas per experimental time. Released LDH is reported as percentage. Data shown are the mean (±SD) of three separate experiments. Zero time LDH release (%): 73.06 * Day 3: 10 µg/mL GO-coated Laminas (GO10) vs. control (CTRL1) $p < 0.05$; * Day 7: APTES-treated, 5 µg/mL GO-coated (GO5) and 10 µg/mL GO-coated Laminas (GO10) vs. control (CTRL1) $p < 0.001$; control Laminas, APTES-treated Laminas vs. 5 µg/mL GO-coated Laminas (GO5) $p < 0.001$; control Laminas, APTES-treated Laminas vs. 10 µg/mL GO-coated Laminas (GO10) $p < 0.001$; * Day 14, day 28: APTES-treated, GO10 Laminas vs. control (CTRL1) Laminas $p < 0.005$.

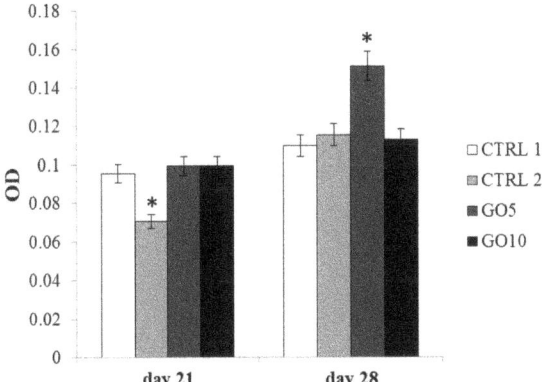

Figure 8. The histogram represents optical density (OD) values of solubilized calcium deposits (orange-red stained) obtained after Alizarin Red staining on bare Laminas (CTRL1), APTES-treated (CTRL2), 5 µg/mL GO-coated (GO5) and 10 µg/mL GO-coated (GO10) Laminas. Twenty Laminas were used for each experimental point, ten Laminas per experimental time. Data shown are the mean (±SD) of three separate experiments. * Day 21: APTES-treated Laminas vs. control (CTRL1) Laminas $p < 0.005$; * Day 28: control Laminas, APTES-treated, 10 µg/mL GO-coated Laminas (GO10) vs. 5 µg/mL GO Laminas (GO5) $p < 0.005$.

4. Discussion

Laminas were covalently enriched with GO, by using APTES as the linker between the Lamina and the GO sheets. This type of functionalization was chosen in order to create on the biomaterial a layer of graphene oxide covalently bound to the scaffold. As a matter of fact, in a previous study [12], porcine bone granules, enriched with GO by exploiting simple physical deposition, were implanted in animals for three months and excess GO was detected in the form of GO aggregates in both hard and soft tissues.

TGA did not allow us to properly quantify the amount of GO functionalized onto Laminas, because the amount of GO was very low and the two materials, cortical membrane and GO, evidenced a weight loss at similar temperatures. An approximately 1.3% GO could be calculated at least for the more concentrated 10 µg/mL sample.

Nevertheless, the GO demonstrated good distribution, through AFM and SEM analyses, on the Lamina and changed completely the appearance of the surface of the cortical membrane. Despite SEM appearance, the enrichment with GO rendered the surface more rough, as already recently evidenced in the case of GO-enrichment of porcine bone granules and collagen membranes [12,13]. Indeed, SEM is not the best technique in order to discriminate GO coverage percentage, but SEM images highlight the formation of layered GO over the Lamina surface (Figure 4). AFM measurements instead evidenced that GO-enriched Laminas are rougher than bare membranes and present a much wrinkled structure (compare panels B, E, and H and C, F, and I in Figure 3).

In particular, the calculated roughness indexes, Rq and Ra, indicate that the non-coated membrane is characterized by lower peaks and therefore a lower roughness as compared to the GO-coated samples, as confirmed by Rmax values. The surface indexes, Sdr and Sdq, confirmed Ra, Rq, and Rmax data because they evidenced a surface enlargement induced by the presence of GO with more and steep peaks, respectively. Nevertheless no significant differences were highlighted between the samples enriched with 5 or 10 µg/mL GO, likely due to a saturation-like effect of the surface with the lowest concentration of GO investigated.

The measured Young's elastic modulus, obtained as an average value calculated from 20–25 force curves in samples of 10 µm × 10 µm dimensions, is 0.77 ± 0.46, 0.83 ± 0.62, and 1.00 ± 0.27 GPa in bare cortical membrane, 5 µg/mL and 10 µg/mL GO-coated membranes, respectively (See Supporting Information, Figures S2–S7). Therefore it does not change very much on GO enrichment, although a small increase could be monitored on increasing the concentration of GO, indicating that GO contributes to matrix stiffening. It is interesting to note that the SD is very high (60%) in the commercial cortical membrane due to the presence of pores and defects, keeps a high value for 5 µg/mL coated Laminas but reduces in Laminas enriched with 10 µg/mL GO, thus highlighting the presence of a homogeneous GO layer in the latter sample. These values are of the same order of those of polyethylene (1.5–2 GPa) and polystyrene (3–3.5 GPa), substrates which have been previously demonstrated to be ideal substrates for the growth of stem cells.

By considering that mesenchymal stem cells demonstrated [18] to sense matrix elasticity and preferentially differentiate depending of the stiffness of the substrate, such values, indicative of stiff matrices, appear proper to favor expression of an osteogenic lineage.

Laminas coated with 5 µg/mL and 10 µg/mL GO were then tested in an in vitro model, by seeding and culturing DPSCs on Laminas surface, in order to evaluate the biocompatibility of GO-enriched Laminas, in terms of cell viability, cytotoxicity, and mineralized matrix deposition.

During the DSPCs differentiation, GO enrichment positively modifies the biological parameters evaluated, thus indicating a good tolerability and an improved biocompatibility. Indeed, GO enrichment, both at 5 and 10 µg/mL concentration, improves the cell spread throughout the surface of the biomaterial thus allowing to hypothesize that GO enrichment is able to promote the adhesion process and favor the formation of a uniform cell layer (See Figure 5). Moreover, GO coating enhanced the growth rate of DPSCs. In fact, cells seeded on GO-coated Laminas after 7 days of culture show cell viability values two-fold that of bare Laminas evidencing confluence after 14 days of culture (see Figure 6). It is worth noting that the above mentioned GO-induced cell viability value is not related to GO concentration as both samples evidence the same effect. Nevertheless it is important to stress that these similarities can be explained by the above mentioned saturation-like effect, with GO covering almost completely the Lamina surface already at the lowest investigated concentration. These results may be associated to the capacity of GO to favor protein adsorption [19]. Indeed, it has been demonstrated that serum proteins absorb quickly and spontaneously to graphene oxide surface to form a corona complex [20] and this adsorption, that demonstrate to be selective for different proteins, may affect adhesion, proliferation, and/or osteogenic differentiation of stem cells [19]. It was also demonstrated that induction of human mesenchymal stem cells (hMSC) differentiation towards different tissue lineages depended on the degree of π-π stacking with graphene and hydrogen bonding as well as electrostatic interactions with GO [19]. In the present case, a positive effect towards adhesion and viability of DPSCs may be induced also by the highly wrinkled surface associated to a small increase in stiffness obtained on GO enrichment. A similar evidence has been already demonstrated for highly convoluted methacrylate-functionalized GO membrane [21] that favored spontaneous stem cell differentiation towards bone lineage even in the absence of osteogenic growth factors.

The cytotoxicity level also appeared significantly reduced for GO-enriched Laminas during 28 days of culture and actually a statistical significant increase of membrane integrity was detected for Laminas enriched with 10 μg/mL GO at all the investigated times. It is worth noting that a slightly different behavior characterizes 5 μg/mL GO-coated Laminas, with cytotoxicity slightly increasing and reaching that of the bare Laminas at 14 and 28 days of culture. These results show that, although GO functionalization demonstrated to promote favorable biological effects on DSPCs, it is necessary to carefully tune the concentration of GO bound to Laminas in order to reach the best compromise of effectiveness and biocompatibility. Indeed, different studies evidenced that doses, as well as size, is a fundamental parameter to consider in order to fully characterize GO toxicity [22]. On the other hand, a relatively high toxicity was detected for APTES-treated Laminas. Despite an APTES-treated different material, such as nanoparticles, did not show any toxic effect on cell membrane integrity [23], polyamines demonstrated [24] to promote leakage of liposomal content from 1-palmitoyl-2-oleoyl-sn-glycero-3-phosphocholine (POPC) liposomes due to interactions of primary ammonium groups with phospholipidic head groups. Similarly APTES-treated Laminas could promote analogous ammonium-phospholipidic headgroups interactions, thus explaining the sustained LDH leakage from cell membrane and the chronic cytotoxicity responsible for the lower proliferation rate as compared to cells seeded on the other investigated samples. These data are very interesting because they highlight that, after aminosilane-functionalization, the subsequent treatment with GO allows to override the negative effect evidenced in the presence of APTES-treated Laminas on cell cytotoxicity.

These results are further supported by mineralized bone matrix deposition which appears increased after 28 days of culture and therefore at the end of the osteogenic differentiation [25] but only for DPSCs cultured on 5 μg/mL GO-coated Laminas. They highlight the important role of GO as responsible for a faster and more intense promotion of bone matrix deposition. To conclude, this is a preliminary study on biocompatibility and lack of cytotoxicity of the GO-functionalized Laminas and further investigations are needed in order to fully characterize their biological properties. First of all a detailed investigation focused on the tuning of GO concentration needed for the covalent functionalization of Laminas will allow to optimize the risk-to-benefit balance and clarify all the factors affected by GO enrichment.

5. Conclusions

This study demonstrated that the relatively homogeneous coating of investigated commercial cortical membranes with GO was relatively easy to obtain. It favored the proliferation rate of DPSCs probably due to the capacity of GO to adsorb proteins present in the medium. Clear evidences of reduced toxicity were evidenced and Laminas enriched with GO 5 μg/mL demonstrated a statistical significant increase of calcium phosphate deposition. The present study is particularly promising and we believe that this material holds potential as useful substrate to facilitate in vivo bone regeneration. Nevertheless it highlights the need to further investigate GO-coated samples in order to tune the concentration of GO that demonstrates the best osteogenic activity and biocompatibility.

Supplementary Materials: The following are available online at http://www.mdpi.com/2079-4991/9/4/604/s1, Figure S1: DPSCs observed with a light microscope before detachment and seeding for osteoblastic differentiation, Figures S2–S7: Original AFM micrographs and representative force curves of pure Lamina, Lamina enriched with 5 µg/mL GO and 10 µg/mL GO used for the mechanical studies, Figures S8–S10: Original AFM micrographs used for the roughness index calculation of pure Lamina, Lamina enriched with 5 µg/mL GO and 10 µg/mL GO, Tables S1–S3: Roughness indexes for pure Lamina, Lamina enriched with 5 µg/mL GO and 10 µg/mL GO.

Author Contributions: R.D.C. enriched the Laminas with GO, S.Z. performed biological analyses, A.V. performed AFM measurements, G.S. and G.I. data curation and supervision, T.D.R. performed TGA analyses, A.C. and A.F. conceptualization, writing original draft, and supervision.

Funding: This work was carried out with the financial support from the University 'G. d'Annunzio' of Chieti-Pescara and MIUR.

Conflicts of Interest: The authors declare no conflict of interest.

References

1. Osteobiol by Tecnoss. Available online: https://www.osteobiol.com (accessed on 15 December 2018).
2. Rossi, R.; Rancitelli, D.; Poli, P.P.; Rasia Dal Polo, R.; Nannmark, U.; Maiorana, C. The use of collagenated porcine cortical lamina in the reconstruction of alveolar ridge defects. A clinical and histological study. *Minerva Stomatol.* **2016**, *65*, 257–268.
3. Jianguo, L.; Yanqun, L.; Haiyan, L.; Yufen, X.; Yongxiang, Z.; Jingxian, L.; Jianping, W. Preparation, bioactivity and mechanism of nano-hydroxyapatite/sodium alginate/chitosan bone repair material. *J. Appl. Biomater. Funct. Mater.* **2018**, *16*, 28–35.
4. Enayati-Jazi, M.; Solati-Hashjin, M.; Nemati, A.; Bakhshi, F. Synthesis and characterization of hydroxyapatite/titania nanocomposites using in situ precipitation technique. *Superlattices Microstruct.* **2012**, *51*, 877–885. [CrossRef]
5. Kealley, C.; Elcombe, M.; van Riessen, A.; Ben-Nissan, B. Development of carbon nanotube reinforced hydroxyapatite bioceramics. *Physic B* **2006**, *385–386*, 496–498. [CrossRef]
6. Novoselov, K.S.; Fal'ko, V.I.; Colombo, L.; Gellert, P.R.; Schwab, M.G.; Kim, K. A roadmap for graphene. *Nature* **2012**, *490*, 192–200. [CrossRef]
7. Kenry, L.W.; Loh, K.P.; Lim, C.T. When stem cells meet graphene: Opportunities and challenges in regenerative medicine. *Biomaterials* **2018**, *155*, 236–250. [CrossRef] [PubMed]
8. Balandin, A.A. Thermal properties of graphene and nanostructured carbon materials. *Nat. Mater.* **2011**, *10*, 569–581. [CrossRef]
9. Wang, X.L. Proposal for a new class of materials: Spin gapless semiconductors. *Phys. Rev. Lett.* **2008**, *100*, 156404. [CrossRef]
10. Guazzo, R.; Gardin, C.; Bellin, G.; Sbricoli, L.; Ferroni, L.; Ludovichetti, F.S.; Piattelli, A.; Antoniac, I.; Bressan, E.; Zavan, B. Graphene-Based Nanomaterials for Tissue Engineering in the Dental Field. *Nanomaterials* **2018**, *8*, 349. [CrossRef]
11. Bresson, E.; Ferroni, L.; Gardin, C.; Sbricoli, L.; Gobbato, L.; Ludovichetti, F.S.; Tocco, I.; Carraro, A.; Piattelli, A.; Zavan, B. Graphene based scaffolds effects on stem cells commitment. *J. Transl. Med.* **2014**, *12*, 296. [CrossRef] [PubMed]
12. Ettorre, V.; De Marco, P.; Zara, S.; Perrotti, V.; Scarano, A.; Di Crescenzo, A.; Petrini, M.; Hadad, C.; Bosco, D.; Zavan, B.; et al. In vitro and in vivo characterization of graphene oxide coated porcine bone granules. *Carbon* **2016**, *103*, 291–298. [CrossRef]
13. De Marco, P.; Zara, S.; De Colli, M.; Radunovic, M.; Lazović, V.; Ettorre, V.; Di Crescenzo, A.; Piattelli, A.; Cataldi, A.; Fontana, A. Graphene oxide improves the biocompatibility of collagen membranes in an in vitro model of human primary gingival fibroblasts. *Biomed. Mater.* **2017**, *12*, 055005. [CrossRef]
14. Radunovic, M.; De Colli, M.; De Marco, P.; Di Nisio, C.; Fontana, A.; Piattelli, A.; Cataldi, A.; Zara, S. Graphene oxide enrichment of collagen membranes improves DPSCs differentiation and controls inflammation occurrence. *J. Biomed. Mater. Res. Part A* **2017**, *105*, 2312–2320. [CrossRef]

15. De Colli, M.; Radunovic, M.; Zizzari, V.L.; di Giacomo, V.; Di Nisio, C.; Piattelli, A.; Calvo Guirado, J.L.; Zavan, B.; Cataldi, A.; Zara, S. Osteoblastic differentiating potential of dental pulp stem cells in vitro cultured on a chemically modified microrough titanium surface. *Dent. Mater. J.* **2018**, *37*, 197–205. [CrossRef]
16. Guo, T.; Cao, G.; Li, Y.; Zhang, Z.; Nör, J.E.; Clarkson, B.H.; Liu, J. Signals in Stem Cell Differentiation on Fluorapatite-Modified Scaffolds. *J. Dent. Res.* **2018**, *97*, 1331–1338. [CrossRef]
17. Xie, H.; Chua, M.; Islam, I.; Bentini, R.; Cao, T.; Viana-Gomes, J.C.; Castro Neto, A.H.; Rosa, V. CVD-grown monolayer graphene induces osteogenic but not odontoblastic differentiation of dental pulp stem cells. *Dent. Mater.* **2017**, *33*, e13–e21. [CrossRef]
18. Engler, A.J.; Sen, S.; Sweeney, H.L.; Discher, D.E. Matrix elasticity directs stem cell lineage specification. *Cell* **2006**, *126*, 677–689. [CrossRef]
19. Lee, W.C.; Lim, C.H.Y.X.; Shi, H.; Tang, L.A.L.; Wang, Y.; Lim, C.T.; Loh, K.P. Origin of Enhanced Stem Cell Growth and Differentiation on Graphene and Graphene Oxide. *ACS Nano* **2011**, *5*, 7334–7341. [CrossRef]
20. Wei, X.-Q.; Hao, L.-Y.; Shao, X.-R.; Zhang, Q.; Jia, X.-Q.; Zhang, Z.-R.; Lin, Y.-F.; Peng, Q. Insight into the Interaction of Graphene Oxide with Serum Proteins and the Impact of the Degree of Reduction and Concentration. *ACS Appl. Mater. Interfaces* **2015**, *7*, 13367–13374. [CrossRef]
21. Tang, L.A.; Lee, W.C.; Shi, H.; Wong, E.Y.; Sadovoy, A.; Gorelik, S.; Hobley, J.; Lim, C.T.; Loh, K.P. Highly wrinkled cross-linked graphene oxide membranes for biological and charge-storage applications. *Small* **2012**, *8*, 423–443. [CrossRef]
22. Alberto, B. Graphene: Safe or Toxic? The Two Faces of the Medal. *Angew. Chem.-Int. Ed.* **2013**, *52*, 4986–4997.
23. Malvindi, M.A.; De Matteis, V.; Galeone, A.; Brunetti, V.; Anyfantis, G.C.; Athanassiou, A.; Cingolani, R.; Pompa, P.P. Toxicity Assessment of Silica Coated Iron Oxide Nanoparticles and Biocompatibility Improvement by Surface Engineering. *PLoS ONE* **2014**, *9*, e85835. [CrossRef]
24. Palermo, E.F.; Lee, D.-K.; Ramamoorthy, A.; Kuroda, K. Role of Cationic Group Structure in Membrane Binding and Disruption by Amphiphilic Copolymers. *J. Phys. Chem. B* **2011**, *115*, 366–375. [CrossRef]
25. Riccio, M.; Resca, E.; Maraldi, T.; Pisciotta, A.; Ferrari, A.; Bruzzesi, G.; De Pol, A. Human dental pulp stem cells produce mineralized matrix in 2D and 3D cultures. *Eur. J. Histochem.* **2010**, *54*, 205–213. [CrossRef]

© 2019 by the authors. Licensee MDPI, Basel, Switzerland. This article is an open access article distributed under the terms and conditions of the Creative Commons Attribution (CC BY) license (http://creativecommons.org/licenses/by/4.0/).

Article

Polymer-Based Graphene Derivatives and Microwave-Assisted Silver Nanoparticles Decoration as a Potential Antibacterial Agent

Angelo Nicosia [1,*], Fabiana Vento [1], Anna Lucia Pellegrino [1], Vaclav Ranc [2], Anna Piperno [3], Antonino Mazzaglia [4] and Placido Mineo [1,5,6,*]

1. Department of Chemical Sciences, University of Catania, V.le A. Doria 6, 95125 Catania, Italy; fabiana.vento@phd.unict.it (F.V.); annalucia.pellegrino@unict.it (A.L.P.)
2. Regional Centre of Advanced Technologies and Materials, Palacký University Olomouc, Šlechtitelů 11, 78371 Olomouc, Czech Republic; vaclav.ranc@upol.cz
3. Department of Chemical, Biological, Pharmaceutical and Environmental Sciences, University of Messina, V.le F. Stagno d'Alcontres 31, 98166 Messina, Italy; apiperno@unime.it
4. CNR-ISMN, Istituto per lo Studio dei Materiali Nanostrutturati, V. le F. Stagno d'Alcontres 31, 98166 Messina, Italy; antonino.mazzaglia@cnr.it
5. Institute for Chemical and Physical Processes CNR-IPCF, Viale F. Stagno d'Alcontres 37, 98158 Messina, Italy
6. Institute of Polymers, Composites and Biomaterials CNR-IPCB, Via P. Gaifami 18, 95126 Catania, Italy
* Correspondence: angelo.nicosia@unict.it (A.N.); placido.mineo@unict.it (P.M.)

Received: 7 October 2020; Accepted: 11 November 2020; Published: 16 November 2020

Abstract: Nanocomposites obtained by the decoration of graphene-based materials with silver nanoparticles (AgNPs) have received increasing attention owing to their antimicrobial activity. However, the complex synthetic methods for their preparation have limited practical applications. This study aims to synthesize novel NanoHybrid Systems based on graphene, polymer, and AgNPs (namely, NanoHy-GPS) through an easy microwave irradiation approach free of reductants and surfactants. The polymer plays a crucial role, as it assures the coating layer/substrate compatibility making the platform easily adaptable for a specific substrate. AgNPs' loading (from 5% to 87%) can be tuned by the amount of Silver salt used during the microwave-assisted reaction, obtaining spherical AgNPs with average sizes of 5–12 nm homogeneously distributed on a polymer-graphene nanosystem. Interestingly, microwave irradiation partially restored the graphene sp^2 network without damage of ester bonds. The structure, morphology, and chemical composition of NanoHy-GPS and its subunits were characterized by means of UV-vis spectroscopy, thermal analysis, differential light scattering (DLS), Field Emission Scanning Electron Microscopy (FE-SEM), Energy Dispersive X-ray analysis (EDX), Atomic Force Microscopy (AFM), and High-Resolution Transmission Electron Microscopy (HRTEM) techniques. A preliminary qualitative empirical assay against the typical bacterial load on common hand-contacted surfaces has been performed to assess the antibacterial properties of NanoHy-GPS, evidencing a significative reduction of bacterial colonies spreading.

Keywords: NanoHy-GPS; antibacterial nanosystems; one-pot microwave-assisted reaction; graphene oxide; silver nanoparticles; polyvinyl alcohol

1. Introduction

The properties of the polymer-based materials have led to their ubiquitous application as structural material not only for common-use objects but also for value-added devices. As examples, these materials are employed in manufacturing of kids' toys, but also to produce biomedical devices such as catheters, ureteral stents, and prosthesis. Especially in the biomedical field, severe infections could occur using invasive devices, due to bacterial contaminations. Besides the necessity to sterilize the materials before their use [1], antimicrobial agents are needed to provide long-term antibacterial efficacy [2]. With this aim, low-weight organic molecules are usually applied as antimicrobial agents and used through spray coating techniques or blended into the polymer matrix during its processing [3,4].

If any bactericidal agent is applied onto the surface of a material, the bacterial adhesion [5] comes in succession due to the reproduction and the formation of colonies, which develop in biofilms (a secretion of exopolysaccharides) [6,7], a protective agent against bactericidal and bacteriostatic substances [8–12].

Viruses also could take advantages from the biofilm, exploiting it as a shield from the environmental stresses, allowing the contamination through biofilm spreading [13]. Such an occurrence represents a huge issue, especially in the latest Severe Acute Respiratory Syndrome CoronaVirus-2 (SARS-CoV-2) pandemic context. To limit the contagion possibilities, the demand of redox-based disinfectant agents to sanitize surfaces has seen an exponential rise, resulting in toxic side-effects towards the environment and wildlife [14].

Moreover, the low adhesiveness, the ubiquitous use of organic-based antibacterial agents, and the subsequential release into the environment has caused direct exposure to life forms, resulting in bioaccumulation for several species worldwide, including humans [15,16], and acting as endocrine disruptors [17]. For these reasons, some of these molecules have also been banned from both European and American health institutions [18]. Nevertheless, the ban concerned only some application fields—these additives are still used, especially as biomedical devices coatings (i.e., surgical suture wires), because of their efficiency towards multiple bacterial targets [18].

An alternative approach to overtaking such a huge issue is represented by surfaces and/or materials having intrinsic long-term antibacterial properties.

In this landscape, a potential solution could be antibacterial polymer-based coatings [19]. Polymers can assure all the required features such as easiness of synthesis and application, long-term stability in environmental conditions, and absence of any degradation product or toxic product leaking. Moreover, antimicrobial agents [19,20] could be loaded in the polymer matrix, performing their long-term activity towards pathogens.

In the field of nanotechnology, Silver NanoParticles (AgNPs) exhibit a broad-spectrum antibacterial activity, against Gram-positive and Gram-negative micro-organisms [21–24], and also multidrug-resistant bacteria [25]. The antibacterial efficiency is attributed to a multifaceted mechanism lying on the release of silver ions [26–29]. The continuous increase of market products containing silver nanoparticles raises the issue of the risks associated with their release in the environment and on the consequent negative effects on human health [30–32].

Graphene Oxide (GO) is a versatile material made up of mono- or few-layers carbon honeycomb structure functionalized with oxidized species (alcohols, epoxides, carboxylic acids), proposed in many application fields. Engineered GO-based materials, due to their biocompatibility, were also proposed as drug delivery systems and for nanomedicine applications [33–36].

Controversial literature data have been reported about the antimicrobial properties of GO; however, its dual oxidative and membrane stress effect has been proven [37–40]. It was verified that polymer coatings containing a suitable amount of GO could prevent the metal substrate from oxidation and bacterial adhesion, while maintaining a positive cell adhesion and response exploitable for surgery implants [41].

The unique features of graphene [37,39] and the antimicrobial properties of the AgNPs could be merged into a complex hybrid system, showing enhanced synergistic antimicrobial effect than the single moiety or their blend [40,42,43]. Moreover, the anchoring of AgNPs onto the graphene platform reduces the risk of aggregation [44–46].

In this framework, our interest has been addressed on the development of a fine-tuned hybrid system combining the properties of polymer, graphene, and AgNPs as a potential on-demand antimicrobial multisurface coating system.

In this system, polymer plays a crucial role since it assures the coating layer/substrate compatibility, making the platform easily adaptable for a specific substrate by changing the polymer of the hybrid system.

In order to ensure the interchangeability of the polymer moiety, the GO functionalization was performed by a synthetic strategy suitable for several polymers. Here, PolyVinyl Alcohol (PVA) was selected as a model moiety and used to produce a PVA@GO covalent system.

Finally, employing a microwave-assisted method to perform a one-pot reaction concerning the simultaneous reduction of AgNPs and GO, nanohybrid systems were obtained (here called PVA@rGO-AgX). The structure, morphology, and chemical composition of PVA@rGO-AgX and its subunits were characterized by means of UV-vis spectroscopy, thermal analysis, differential light scattering (DLS), Field Emission Scanning Electron Microscopy (FE-SEM), Energy Dispersive X-ray analysis (EDX), Atomic Force Microscopy (AFM), and High-Resolution Transmission Electron Microscopy (HRTEM) techniques. Finally, preliminary biological tests have been performed as a proof of concept for its antibacterial activity.

The synthetic pathway exposed here allows fine-tuning of the nanosystem features, thus representing the first approach towards the synthesis of a novel class of NanoHybrid Systems based on Graphene, Polymer, and Silver, namely, NanoHy-GPS.

To our knowledge, an on-demand polymer antibacterial coating nanosystem has not been found yet.

2. Materials and Methods

2.1. Synthesis

Natural graphite powder (diameter 5–10 μm, thickness 4–20 nm, layers < 30, purity > 99.5 wt.%), polyvinyl alcohol, silver nitrate, and all the other reagents and solvents used in this work were purchased by Sigma-Aldrich (Merck Group, Milan, Italy).

2.1.1. GO Synthesis

Graphene oxide (GO) was prepared by oxidizing graphite powders (diameter 5–10 μm, thickness 4–20 nm, layers < 30) according to our previously reported procedure [34] using the Hummers method [47]. Briefly, graphite (2 g) and concentrated sulfuric acid (350 mL) mixture was cooled at 0 °C under stirring. Then, sodium nitrate (1 g) and potassium permanganate (8 g) were slowly added, the temperature was raised up to 40 °C, and the mixture stirred for 1 h. Deionized water (250 mL) was slowly added into the solution (determining an increase of temperature up to 70 °C), the temperature was raised up to 98 °C, and the mixture was stirred for 30 min. Finally, 52 mL of H_2O_2 (30%) was added and the bright yellow suspension was filtered by using a Millipore Membrane (0.1 μm) under vacuum and washed with HCl (4%) and water to reach a neutral pH. The solid was dried to obtain the graphite oxide as a brown powder (1.8 g). Aqueous suspension of graphite oxide (500 mg in 35 mL of water) was exfoliated by ultrasonication (40% W, 8 h) using a UW 2070 SONOPLUS, Bandelin Electronic (Berlin, Germany). The dispersion was diluted with deionized water and centrifugated (10,000 rpm for 12 min). The GO supernatant was dried to recovery GO powder.

2.1.2. Polymer-GO Covalent Adduct Synthesis

The Graphene-Oxide–Polyvinyl-alcohol esterification (PVA@GO) was conducted through a slightly modified method developed by Salavagione et al. [48]. Briefly, 20 mg of GO and 200 mg of PolyVinyl Alcohol (PVA) were dissolved in dimethyl sulfoxide (DMSO, 10 mL) at 70 °C under stirring in nitrogen atmosphere. After 24 h, the mixture was cooled at room temperature. Then, N,N'-Dicyclohexylcarbodiimide (DCC, 926 mg, 4.5 mmol) and N,N-Dimethylpyridin-4-amine (DMAP, 69 mg, 0.56 mmol), previously solubilized in DMSO (10 mL) in nitrogen atmosphere, were added to the PVA solution. The reaction was kept under stirring in nitrogen atmosphere for 3 days;

after that, the mixture was precipitated in methanol (50 mL) and centrifuged (9000 rpm, 20 min). The precipitate was dispersed in 70 mL hot water (70 °C). So, the water solution, containing PVA@GO, was concentrated by means of a rotavapor, and then coagulated in methanol and centrifugated (9000 rpm, 20 min). The procedure was repeated three times. The precipitate was dried for 24 h in oven (50 °C) under vacuum.

2.1.3. Microwave-Assisted Silver Nanoparticles Synthesis

In order to reduce the silver nanoparticles on the GO platform, a microwave-assisted reaction was conducted on PVA@GO aqueous solution (0.5 mg/mL). The solution was put in a 10 mL vessel and proper amounts of Silver Nitrate ($AgNO_3$) were added. Once the solution was stirred, a suitable amount (5 mL) of Dimethylformamide (DMF) was added and the mixture was sonicated for 2 min. The vessel was capped and inserted into the microwave holder. The reaction was conducted at a fixed power of 300 W for two minutes, cooling the system by means of air flux.

The total reaction mixture was mixed with methanol (20 mL) and concentrated using a rotary evaporator (60 °C and vacuum). Then, the mixture was precipitated in ethyl ether (25 mL), the solid was separated through centrifugation (9000 rpm, 20 min), then dried in a vacuum oven at 50 °C overnight.

By increasing the $AgNO_3$ amount (35.7 µg (0.21 µmol), 0.33 mg (1.96 µmol), and 3 mg (18.7 µmol)) added during the synthetic procedure, three products were obtained, namely, PVA@rGO-Ag1, PVA@rGO-Ag2, and PVA@rGO-Ag3, respectively.

For sake of comparison, by means of the same microwave-assisted procedure, without using the $AgNO_3$, samples of reduced PVA@GO (called PVA@rGO) and a sample of reduced GO (called rGO) were produced as well.

2.1.4. Sodium-Borohydride-Mediated Silver Nanoparticles Synthesis

Besides, Silver Nanoparticles (AgNPs) have been also synthetized through the typical chemical reduction of silver nitrate in aqueous solution [49]. Briefly, 750 mL of sodium borohydride water solution (2 mM) were prepared and left under high-speed stirring. Then, 250 mL of silver nitrate water solution (1 mM) was added through a dropping funnel. The reaction was left under high stirring until the end of the silver nitrate solution. In order to stabilize the suspension and have certainty about the reduction of the whole silver salt added, the mixture was stirred for an additional 30 min. The obtained AgNPs suspension, stored at 5 °C, is stable for months. The Localized Surface Plasmon Resonance (LSPR) signal at 394 nm of such AgNPs has been checked by UV-vis measurements.

2.1.5. Preliminary Antibacterial Tests

Bacterial population from common hand-contacted surfaces have been recovered and streaked [50] over the Plate Count Agar (PCA). In detail, sterile swabs, premoistened in Maximum Recovery Diluent (MRD), were streaked over common hand-contacted surfaces (i.e., door handles and handrails) with size 100 cm^2. In order to ensure a good capture of bacteria, the swab constantly rotated and uniformly swiped in all directions of the tested surface. Then, the microorganisms transferred to the swab were released into 10 mL of MRD, transported in the microbiology laboratory (about 5 min), and used to test the several compounds, namely, AgNPs, PVA@rGO-Ag2, PVA@rGO-Ag1, $AgNO_3$, and PVA@GO. Each compound (100 µL, 27 µg/mL in Ag content and/or 142 µg/mL in PVA@GO content) was previously deposited onto the PCA. Then, bacteria in MRD (100 µL) were spread on treated PCA and incubated for 24 h at 37 °C. Each biological experiment was performed in triplicate. As control, bacteria in MRD without any compound was used. Microbial growth was evaluated according to the Colony-Forming Unit (CFU) assay, assuming that each colony has originated from a single bacterium [51]. The CFU for cm^2 of sampling surface (CFU/cm^2) value were calculated from Equation (1):

$$CFU/cm^2 = (\text{Number of Colonies} \times \text{Volume of MRD (10 mL)})/(\text{Volume spread on PCA (0.1 mL)} \times \text{Sampled surface (100 } cm^2\text{)}). \tag{1}$$

2.2. Methods

UV-vis spectra were recorded at 25 °C by a Cary 60 UV-vis spectrophotometer (Agilent Technologies, Santa Clara, CA, USA) in quartz cells (1 cm optical path), using water as a solvent. The concentration of the analyzed systems were 0.3 mg/mL for GO, rGO, PVA, PVA@GO, and PVA@rGO; 3.5 mg/L for AgNPs; instead, the PVA@rGO-Ag sample concentration was 30 µg/mL due to the high intensity of the LSPR signal.

^{1}H NMR and COSY spectra (acquired at 27 °C with a spin lock time of 0.5 s) were obtained on a ^{UNITY}INOVA instrument (Varian, Agilent Technologies, Santa Clara, CA, USA) operating at 500 MHz and using VNMR for acquisition and spectra processing. Samples were dissolved in water-d6 and the chemical shifts expressed in ppm by comparison with the water residue signal.

High-pressure microwave-assisted reactions were performed in a single-mode microwave reactor CEM Discover S-Class (CEM Corporation, Matthews, NC, USA) equipped with a calibrated infrared temperature sensor, employing capped sealed pressure-rated vessels (10 mL).

The Dynamic Light Scattering (DLS) measurements were performed by a miniDAWN Treos (Wyatt Technology, Santa Barbara, CA, USA) multiangle light scattering detector, equipped with a Wyatt QELS DLS Module. The measurements were performed at 25 °C using water (LC-MS grade) as a solvent, previously filtered with 0.2 µm filter. Size distributions were obtained using the ASTRA 6.0.1.10 software (Wyatt Technology, Santa Barbara, CA, USA).

Thermogravimetric analyses were performed by means of Pyris TGA7 (Perkin Elmer, Waltham, MA, USA) in the temperature range between 50 and 800 °C, under an air flow of 60 mL min^{-1} and heating rate of 10 °C min^{-1}.

The combined system NTEGRA Spectra (NT-MDT Co., Zelenograd, 124482, Moscow, Russia) was utilized to acquire sample topography and Nova Px ver 3.4.0 rev. 19040 software (NT-MDT Co., Zelenograd, 124482, Moscow, Russia) was used for the data analysis. The surface morphology was obtained by the means of semicontact mode (height and phase) with the NSG30 (High Resolution NONCONTACT "GOLDEN", NT-MDT, Moscow, Russia) cantilever having a force constant of 22–100 N/m and resonant frequency 240–440 kHz, and the ACTA-SS (AppNano, Mountain View, CA, USA) cantilever having a force constant of 13–77 N/m and resonant frequency of 200–400 kHz. The scanning rate was 0.3 Hz. During the measurement, the humidity was in range of 45–55% and the temperature was RT. The height profile was calculated using Gwyddion 2.51 software [52]. The sample was dissolved in water and sonicated for at least 60 min. Total volume of 3 µL of the as-prepared sample was deposited at freshly cleaved MICA substrate via drag and drop method and left to dry.

The morphology was investigated using the field emission scanning electron microscope ZEISS Supra 55 VP (Zeiss, Oberkochen, Germany). The atomic composition of the samples was analyzed through energy dispersive X-ray analysis, using an INCA-Oxford windowless detector, having a resolution of 127 eV at the full-width half-maximum (FWHM) of the Mn Kα.

HRTEM images were obtained using a high-resolution transmission electron microscope HR-TEM FEI Titan G2 60–300 (Thermo Fisher Scientific, Waltham, MA, USA) with an X-FEG type emission gun, operating at 80 kV. This microscope was also supplied with a Cs image corrector and a STEM High-Angle Annular Dark-Field detector (HAADF). For these analyses, a droplet of the material ultrapure H$_2$O dispersion (0.1 mg mL^{-1}) was deposited onto a carbon-coated copper grid and dried.

3. Results

3.1. Preparation of PVA@rGO-Ag Hybrid Systems

The esterification of the GO was performed exploiting the PVA chains hydroxyl side-groups, as described in Scheme 1 (Step 1), using DCC and DMAP as coupling reagents. In order to promote the anchoring of AgNPs on GO platform, the direct reduction microwave-assisted reaction of silver ions was performed (Scheme 1, Step 2). PVA@GO and AgNO$_3$ were well dispersed in water/DMF and subjected to microwave irradiation for 2 min (300 W, temperature below 180 °C under pressurized air cooling). PVA@rGO-Ag1, PVA@rGO-Ag2, and PVA@rGO-Ag3 samples were obtained increasing the AgNO$_3$ amount (see experimental procedure). Samples called PVA@rGO (reduced PVA@GO) and rGO (reduced GO) were obtained by microwave irradiation of GO and PVA@GO and used to investigate the restoring of graphene sp^2 network.

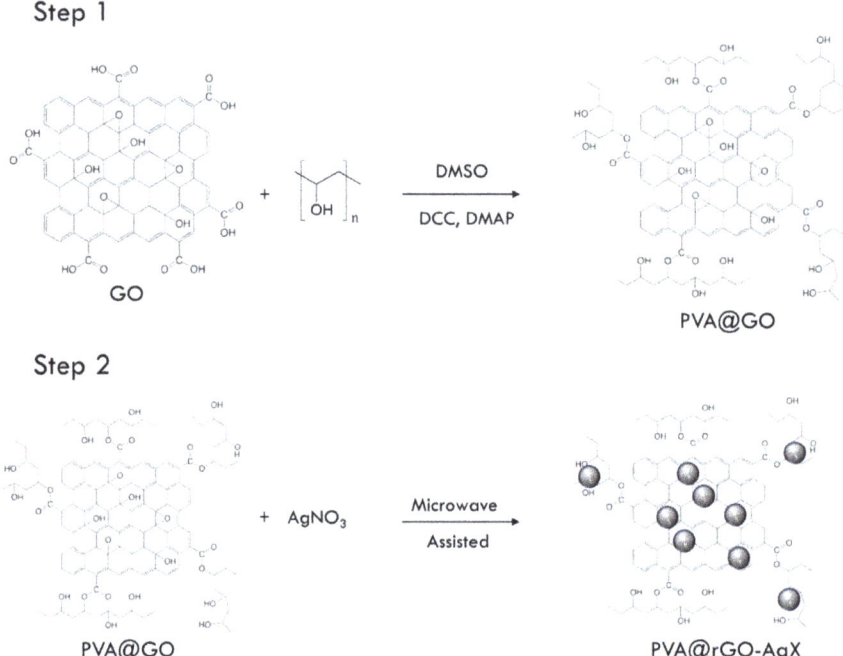

Scheme 1. Schematic representation of PolyVinyl Alcohol-Graphene Oxide-silver hybrid systems (PVA@rGO-AgX) preparation.

3.2. Characterization

The molecular structure of the GO polymer derivative (PVA@GO) was confirmed by ^{1}H-NMR and COSY spectroscopies. The ^{1}H NMR spectrum of the PVA@GO (Figure S1) shows the typical signals of PVA protons at 1.15–1.97 ppm (CH$_2$), and at 3.57–4.17 ppm (CH). Noteworthily, the appearance of the novel signal at 5.05 ppm, attributed to the ester CH protons, proved the successful esterification reaction. The correlations between methylene protons (1.15–1.95 ppm) and CH protons at 5.05 ppm in the COSY-NMR spectrum (Figure S2) further confirms the occurrence of the esterification reaction with a degree of functionalization comparable with literature data (about 2.5%) [48].

A microwave-assisted reaction was set up to decorate the PVA@GO nanosystem with different amounts of AgNPs. The same procedure was applied to GO and PVA@GO in the absence of silver salt (see Experimental Section).

The spectroscopic properties of the nanohybrid systems were investigated in water dispersion through UV-vis spectroscopy (Figure 1). The PVA spectrum (grey dashed line, 0.3 mg/mL) indicates a negligible absorption signal. On the other hand, PVA@GO (red dashed line, 0.3 mg/mL) exhibits a wide scattered signal from 200 nm to 800 nm.

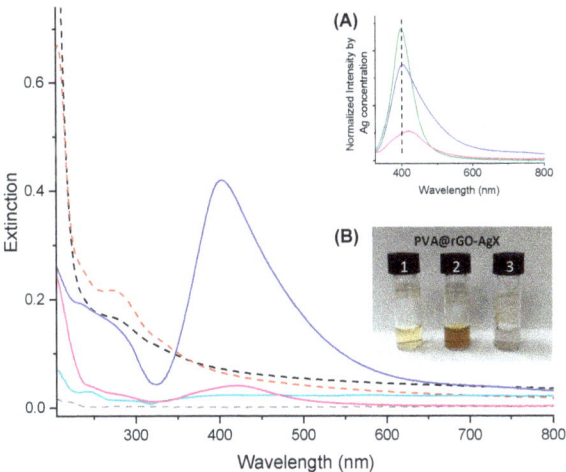

Figure 1. UV-vis spectra of water dispersions/solutions of PolyVinyl Alcohol (PVA) (grey dashed line, 0.3 mg/mL), PVA@GO (red dashed line, 0.3 mg/mL), PVA@rGO (black dashed line, 0.3 mg/mL), PVA@rGO-Ag1 (magenta line, 30 µg/mL), PVA@rGO-Ag2 (blue line, 30 µg/mL), PVA@rGO-Ag3 (cyan line, 30 µg/mL). Inset (**A**): UV-vis spectra normalized by Ag weight content of AgNPs (green line), PVA@rGO-Ag1 (magenta line), PVA@rGO-Ag2 (blue line). Inset (**B**): Digital image of water dispersions (0.3 mg/mL) of PVA@rGO-Ag1 (vial 1), PVA@rGO-Ag2 (vial 2), and PVA@rGO-Ag3 (vial 3).

It is noticeable that the microwave-assisted reaction determines the GO platform reduction of the PVA@GO, obtaining a new system named PVA@rGO (black dashed line, 0.3 mg/mL), as suggested by changes in the absorption curve slope.

To confirm this hypothesis, a sample of GO was treated in microwave-assisted reaction conditions. The red-shift of the π-π* transition peak from 237 nm (attributed to GO, Figure S3) to 266 nm (attributed to rGO, Figure S3), together with a different slope, confirmed that the reduction of GO occurred. The rGO shows also an absorption signal at 205 nm, according to the increase of π-π attraction between rGO platforms [53].

The Localized Surface Plasmon Resonance (LSPR) provides qualitative considerations about the AgNPs deposited onto the hybrid nanosystem.

As expected, UV-vis spectra of the three PVA@rGO-AgX water dispersion (30 µg/mL) show spectroscopic differences due to the different Ag content and to the interactions with the polymer-based substrate. In particular, the PVA@rGO-Ag1 (Figure 1, magenta line) shows the LSPR extinction signal at 420 nm (FWHM about 100 nm), while PVA@rGO-Ag2 (Figure 1, blue line) shows a band centered at 403 nm (FWHM about 100 nm), generally corresponding to AgNPs size of about 50 nm and 15 nm, respectively [54].

To shed light on the substrate influence towards the AgNPs formation, in Figure 1A, we report the UV-vis profile as normalized intensity by Ag concentration of free AgNPs (green line), PVA@rGO-Ag1 and -2 (magenta and blue line, respectively) vs. wavelength. In contrast with UV-vis of the free AgNPs spectrum (showing LSPR at 394 nm, FWHM about 60 nm), broadened and weaker LSPR bands are exhibited by both the PVA@rGO-AgX systems. This variation could be due to the interaction with the polymer-based substrate [55,56], as confirmed by the increase of LSPR red-shift by decreasing the Ag content into the nanohybrid systems.

The different color of the PVA@rGO-Ag1 and -2 water dispersions (respectively, brownish and yellowish; see Figure 1B) is a naked-eye evidence of their quali- quantitative differences.

In the case of PVA@rGO-Ag3, the higher amount of Ag salt in the synthesis procedure determines the formation of agglomerated broad-sized Ag particles, resulting in unstable microscopic suspension (Figure 1, cyan line, and Figure 1B).

The quantitative determination of the Ag on the nanosystems was performed through thermogravimetric analysis (Figure 2) [46,57,58]. The Ag weight (%) content in the different samples has been estimated, obtaining 5%, 16%, and 87% (w/w) of Ag for PVA@rGO-Ag1, -2, and -3, respectively. For the sake of clarity, TGA traces of GO (dotted line) and PVA@GO (black line) are also reported.

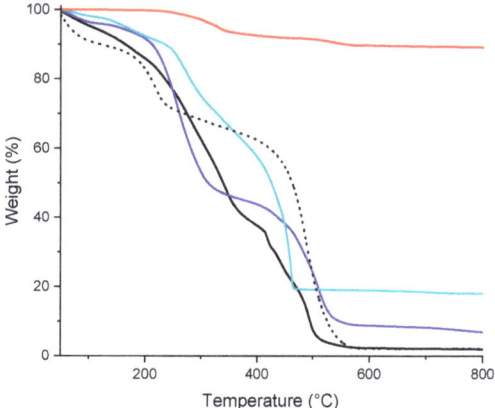

Figure 2. Thermogravimetric analysis (TGA) thermograms of Graphene Oxide (GO) (dotted line), PVA@GO (black line), PVA@rGO-Ag1 (blue line), PVA@rGO-Ag2 (cyan line), and PVA@rGO-Ag3 (red line), air atmosphere.

A morphological and qualitative characterization of the hybrid nanosystems has been conducted employing FE-SEM and EDX investigations (Figure 3). All the PVA@rGO-Ag nanohybrid systems present similar smooth surfaces attributed to PVA@rGO backbone. The AgNPs displayed spherical and quite regular morphology in all three analyzed samples, with a distribution of agglomerates in samples PVA@rGO-Ag2 (Figure 3B) and, in particular, a very large distribution in PVA@rGO-Ag3 (Figure 3A). The EDX analysis confirms the nature of the nanoparticles, as evidenced by a peak at 2.98 keV attributed to Ag Lα line, and shows a strong decrease of Ag atomic percentage from PVA@rGO-Ag3 to PVA@rGO-Ag1 of 83%, 22%, and 0.5% (Figure 3A–C), respectively. In addition, only the signal at 0.28 keV related to the C Kα peak arising from PVA@rGO backbone is present.

The PVA@rGO-Ag1 represents the most reliable sample to investigate the nanohybrid system morphology, because of the lack of AgNPs agglomerates. In fact, despite the absence of AgNPs evidence in the FE-SEM image, the related EDX spectra (Figure 3C) confirms the AgNPs presence, which are well-dispersed and embedded within the PVA@rGO backbone.

PVA@rGO-Ag1 sample was characterized by means of HRTEM and AFM analyses. AFM was used to investigate the localization of Ag nanoparticles on the surface of rGO and to evaluate the topography of synthesized hybrid nanostructures. It is hypothesized that a considerably higher value of Ag concentration would indeed cover the rGO moiety (as evidenced by SEM analysis in Figure 3A). The phase images (Figure 4A,C and related 3D rendering in Figure 4B,D) confirm this and show that the sample is in the form of a PVA homogeneous film containing both AgNPs and rGO platform. The height profile calculations (shown in Figure S5) revealed that the rGO moiety is basically a few layers of aggregates with a height of around 50 nm. Its surface is functionalized by PVA and covered by AgNPs. According to the UV-vis results, the AgNPs size distribution ranged from few to dozens of nanometers.

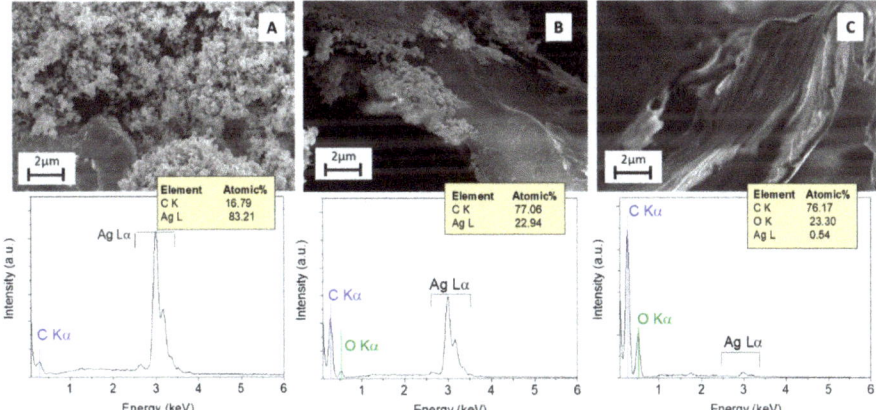

Figure 3. (**A**) Field Emission Scanning Electron Microscopy (FE-SEM) images and the related Energy Dispersive X-ray analysis (EDX) spectra of PVA@rGO-Ag3; (**B**) PVA@rGO-Ag2; (**C**) PVA@rGO-Ag1.

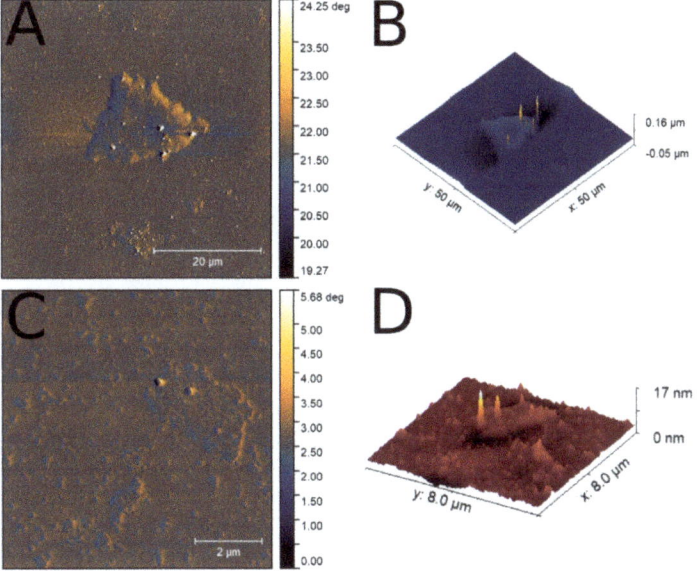

Figure 4. Atomic Force Microscopy (AFM) phase images of PVA@rGO-Ag1 (**A**) 50 × 50 μm and (**C**) 8 × 8 μm and (**B,D**) the related 3D Height.

HRTEM images of PVA@rGO-Ag1 (Figure 5A) show the typical morphology of hybrid nanocomposite materials. Taking into account that the starting GO is produced as folded and large sheets, the effects of the concurrent microwave irradiation and deposition of AgNPs have promoted the unfolding and formation of well-exfoliated graphene sheets having different levels of transparency. As a result, PVA@rGO-Ag1 shows the typical morphology of functionalized reduced graphene (Figure 5B), as observed in our previous papers [34,35], evidencing planar backbone conformation and transparency. The AgNPs were detected as dark spots homogeneously dispersed over the PVA@rGO background material (Figure 5C,D), confirming a random formation of nucleation site within the polymer-based nanosystem. The AgNPs localization, also within the PVA matrix, is in accordance with literature data, since the PVA acts as a capping agent for AgNPs [59–62]. Thus, GO-oxygenated functional groups and PVA -OH groups act as homogeneous NPs stabilizers. Considering the AgNPs' size obtained by HRTEM and UV-vis analyses, the data apparently do not match each other: this ostensible contradiction could be explained by considering the observable properties of the system. Indeed, while HRTEM technique reports the morphology of the system in dry state, the LSPR signal is influenced by the chemical environment of the AgNPs (solvent, PVA -OH groups, graphene, etc.).

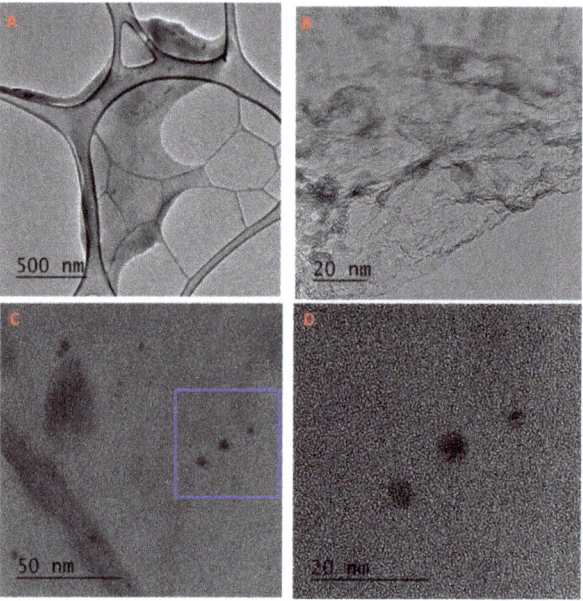

Figure 5. Representative high-resolution transmission electron microscopy (HRTEM) images of PVA@rGO-Ag1 (**A–D**). Area visualized in (**D**) is highlighted in (**C**) by the blue rectangle.

The size of nanosystems in water dispersion were investigated by Dynamic Light Scattering (DLS) analyses. Generally, DLS values are not indicative of the morphology and size of the graphene nanosystems, but provide information about their hydrodynamic size [63,64]. The functionalization of GO with PVA resulted in two hydrodynamic radius distributions (Figure 6, black line) centered at 5 nm and 70 nm: those are similar to that of the pure PVA (grey line), even while having different relative intensities.

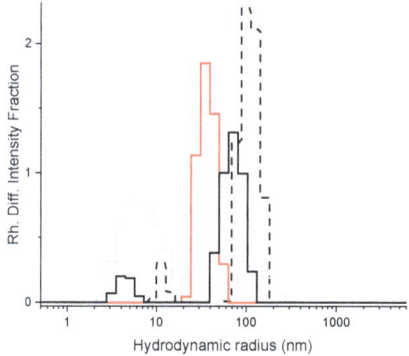

Figure 6. Dynamic Light Scattering (DLS) measurements of GO (black dashed line), PVA (grey line), PVA@GO (black line), and PVA@rGO-Ag1 (red line) in water (samples concentration 12.5 µg/mL).

The microwave-assisted procedure for PVA@rGO-AgX synthesis slightly affects the hydrodynamic radius of the nanohybrid system (about 35 nm) as evidenced by PVA@rGO-Ag1 (Figure 6, red line), suggesting an increase of hydrophilicity due to the silver loading.

3.3. Preliminary Antibacterial Activity Assay of the PVA@rGO-Ag

In order to investigate the antibacterial performances of the PVA@rGO-Ag nanosystem, a qualitatively empirical test has been performed considering the typical bacterial population of common hand-contacted surfaces.

PVA@rGO-Ag2 nanosystem was selected due to the highest AgNPs content because the PVA@rGO-Ag3 does not show clear physicochemical properties of AgNPs (i.e., LSPR signal). The PVA@rGO-Ag2 antibacterial efficacy has been qualitatively compared with the bare AgNPs (the synthetic procedure is described in Materials and Methods), silver nitrate, and PVA@GO (Figure 7).

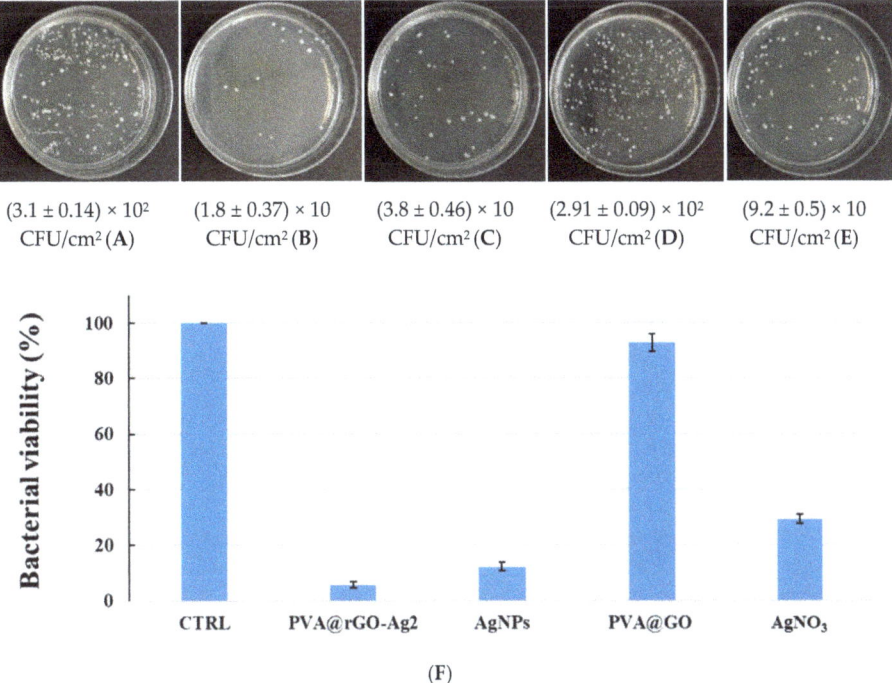

Figure 7. Antibacterial assay in plate count agar after 24 h of incubation: (**A**) Control experiment (CTRL); (**B**) PVA@rGO-Ag2; (**C**) AgNPs; (**D**) PVA@GO; (**E**) AgNO$_3$; (**F**) viability percentage. The deviation standard in CFU/cm^2 has been derived from triplicate test. The CFU for cm^2 of sampling surface (CFU/cm^2) values were calculated from Equation (1).

Control experiment (Figure 7A) and PVA@GO (Figure 7D) Petri dish show a higher bacterial load, in terms of CFU/cm^2, than all other conditions. As expected, several colony morphologies, such as golden-yellow colonies, large, and white-raised colonies were detectable from cultures in PCA, suggesting the presence of multiple bacterial species in the sampling. Moreover, bacterial viability suggests that the effects of bare PVA@GO towards bacterial proliferation resulted to be negligible (Figure 7F). The condition with AgNO$_3$ (Figure 7E) shows a significant reduction of the bacterial load (estimated around 70%, Figure 7F), even if the presence of multiple bacterial species is still maintained. Finally, the conditions with PVA@rGO-Ag2 (Figure 7B) and AgNPs (Figure 7C) show a reduction of

88% and 94% (Figure 7F), with total killing of some bacterial species. The same experiment involving PVA@rGO-Ag1 was performed, but the bacterial viability was not significantly reduced (Figure S6).

4. Discussion

Nanotechnologies could fulfil the need to find an alternative to the common organic molecular systems used to make antimicrobial surfaces. The use of AgNPs embedded within a carbon-based nanosystem could provide a combined effect towards bacterial colonization [40]. Indeed, graphene toxicity provides mechanical and oxidative stresses, while the AgNPs are a well-known broad-spectrum antibacterial agent. Moreover, the polymer functionalization is fundamental to ensure a strong sticking towards a substrate and/or a homogeneous mix within a polymer matrix, thanks to the similar chemical–physical features with target substrates.

On this basis, the new NanoHybrid system composed of Polymer, Graphene, and AgNPs—namely, NanoHy-GPS—was designed and has here been systematically described, employing PVA moiety as a polymer model.

Aiming to develop an interchangeable GO-based system, the GO grafting-to procedure has been chosen to allow the use of any polymer having functionalities (as end- and/or side- groups) suitable for esterification reactions. Thanks to the interchangeability of the polymer moiety, the NanoHy-GPS system represents a new concept of on-demand potentially tunable antimicrobial nanosystem.

The synthesis of PVA@GO was performed through a slightly modified procedure (Scheme 1) than that proposed by Salavagione et al. [48], and the successful of the reaction was confirmed through NMR analyses (^{1}H-NMR and ^{1}H-^{1}H COSY, Figures S1 and S2).

Then, the PVA@GO was used as a platform to be decorated with silver nanoparticles (Scheme 1). With this aim, a one-pot microwave-assisted reduction procedure (simultaneously reducing the GO and the silver ions) has been performed. Three so-called PVA@rGO-AgX nanosystems were produced with increasing Ag contents (PVA@rGO-Ag1, -2, and -3) to investigate the microwave-assisted synthesis efficiency and the antibacterial activity.

The presence of the PVA@GO platform influences the AgNPs formation—probably, the different chemica–physical nature of the nucleation sites ensured by the polymer-functionalized carbon platform, coupled with the role of PVA moiety to act as NPs stabilizer, allowed the growth of AgNPs, as revealed by UV-vis measurements (see Figure 1). As a confirmation, in absence of PVA@GO during the microwave-assisted procedure, the AgNPs formation is hindered (as revealed by the absence of LSPR signal, see Figure S4). The quantitative determinations of the silver content have been performed by TGA technique (see Figure 2), which is a reliable approach to determine the metal content within hybrid systems [57,65]. The Ag weight (%) content in the different samples was calculated, obtaining 5%, 16%, and 87% (w/w) of Ag for PVA@rGO-Ag1, -2, and -3, respectively.

SEM characterization showed the general morphology of the hybrid systems and the AgNPs distribution (see Figure 3), with the smooth and laminar surface of PVA@GO platform and regular and spherical AgNPs. The SEM images displayed that the increase of AgNPs induces an almost complete covering of the PVA@GO substrate and a formation of nanoparticles agglomerates. Conversely, in sample PVA@rGO-Ag1, with lower amount of AgNPs, the nanoparticles are barely visible and may be embedded in the composite (Figure 3C).

EDX analysis showed the decreasing trend of the amount of AgNPs from PVA@rGO-Ag3 to PVA@rGO-Ag1 samples, confirming the trend of the data obtained through TGA. In addition, the presence of any contaminants is excluded.

To better describe the nanohybrid system morphology, the PVA@rGO-Ag1 sample was selected as a model system, because a higher Ag amount (as PVA@rGO-Ag2) would hide the PVA@rGO substrate. The AFM (Figure 4) and HRTEM micrographs (Figure 5) of PVA@rGO-Ag1 evidenced that the nucleation of the AgNPs occurs onto the GO platform and within PVA matrix as well. This data confirmed the exploitation of both the microchemical environment of GO oxygenated groups and the PVA hydroxyl groups interactions with the external surface of silver nanoparticles [66].

Thanks to the polymer-functionalization, the NanoHy-GPS might ensure a long-lasting coating, paving the way to its antibacterial activity expression over time. The NanoHy-GPS is conceptually suitable to be blended within the polymer used in the production of invasive biomedical devices. Moreover, aiming to replace the commonly used antibacterial agents, it could be also sprayed or applied by dipping on any polymer-based surfaces already produced.

With the aim to potentially use the NanoHy-GPS as an antibacterial agent for surfaces of common-use objects, experimental proof of concept was carried out, taking into account the bacterial population of common hand-contacted surfaces. The results revealed that PVA@rGO-Ag2 exhibited improved antibacterial properties than that of bare AgNPs (Figure 7).

Especially in the context of a worldwide spreading pandemic, the features exposed here could also candidate NanoHy-GPS as an antibacterial agent for common-use objects, helping to reduce the spreading and related toxic effects due to the common disinfectants agents [14,67].

Nevertheless, further investigation on antibacterial activity against pathogens involved in various common infections (i.e., Gram (+) and Gram (−) bacterial cells) are in due course, with the aim of optimizing antimicrobial efficacy by tuning the physical–chemical properties of NanoHy-GPS.

5. Conclusions

Here, we reported a flexible synthetic strategy for the fabrication of new NanoHybrid systems (NanoHy-GPS) composed of Polymer, Graphene, and AgNPs. It was shown that the synthetic pathway exploits the microwave irradiation for the simultaneous reduction of silver nanoparticles and GO without the use of reductants and surfactants. AgNPs content in NanoHy-GPS (from 5% to 87%) is quali- quantitatively influenced by the amount of Ag salt used during the microwave-assisted reaction, determining variations in both NPs distribution and size of their agglomerates. The method allows binding of different polymers having suitable active groups (alcohol or amine) and control of the AgNPs content in the nanosystem, giving the capability to tune the interfacial interactions towards targeted substrates (matching their relative chemical nature) and optimizing the homogeneity of the dispersion of the GO-derivatives within specific polymer matrices. The NanoHy-GPS is projected to be a valid alternative towards the common antibacterial agents, which incurs to leaks within the environment and/or within organisms' tissue.

Supplementary Materials: The following are available online at http://www.mdpi.com/2079-4991/10/11/2269/s1, Figure S1: ^{1}H-NMR spectrum of PVA@GO, Figure S2: COSY ^{1}H-NMR spectrum of PVA@GO, Figure S3: Normalized UV-vis spectra of GO and rGO, Figure S4: UV-vis spectra of the AgNO$_3$ solutions before and after the microwave-assisted procedure, Figure S5: AFM profiles and height calculations of PVA@rGO-Ag1, Figure S6: Antibacterial assay of PVA@rGO-Ag1.

Author Contributions: Conceptualization, A.N. and P.M.; methodology, A.N. and P.M.; validation, A.P., A.M., and P.M.; investigation, A.N., A.L.P., F.V., and V.R.; resources, A.N., F.V., A.P., and P.M.; writing—original draft preparation, A.N., and P.M.; writing—review and editing, A.N., F.V., A.L.P., A.P., V.R., A.M., and P.M.; visualization, A.N., F.V., A.P., and P.M.; supervision, P.M.; project administration, A.N. and P.M.; funding acquisition, P.M. All authors have read and agreed to the published version of the manuscript.

Funding: This work was partially supported by the University of Catania (PIAno di inCEntivi per la RIcerca di Ateneo, PIACERI–Linea 2) and by Ministero dell'Istruzione, dell'Università e della Ricerca (MIUR) PRIN Prot. 2017YJMPZN-005.

Acknowledgments: Many thanks to Daniele Mezzina and Lidia Mezzina for their support with the experimental setup. The authors would also like to thank Domenico Franco (Dipartimento di Scienze Chimiche, Biologiche, Farmaceutiche ed Ambientali dell'Università di Messina, Viale F. Stagno d'Alcontres 31, Messina 98166, Italy) for his help with the antibacterial assay.

Conflicts of Interest: The authors declare no conflict of interest.

References

1. Kamal, G.D.; Pfaller, M.A.; Rempe, L.E.; Jebson, P.J.R. Reduced Intravascular Catheter Infection by Antibiotic Bonding. *JAMA* **1991**, 265. [CrossRef]
2. Bowersock, T.L.; Woodyard, L.; Hamilton, A.J.; DeFord, J.A. Inhibition of Staphylococci by vancomycin absorbed on triidodecylmethyl ammonium chloride-coated intravenous catheter. *J. Control. Release* **1994**, *31*, 237–243. [CrossRef]
3. Shrivastava, A. Additives for Plastics. In *Introduction to Plastics Engineering*; William Andrew (Elsevier): Oxford, UK, 2018; pp. 111–141. [CrossRef]
4. McKeen, L. Introduction to Plastics and Polymers. In *The Effect of Sterilization Methods on Plastics and Elastomers*; William Andrew (Elsevier): Oxford, UK, 2018; pp. 41–61.
5. Costerton, J.W. Bacterial Biofilms: A Common Cause of Persistent Infections. *Science* **1999**, *284*, 1318–1322. [CrossRef] [PubMed]
6. Kingshott, P.; Wei, J.; Bagge-Ravn, D.; Gadegaard, N.; Gram, L. Covalent Attachment of Poly(ethylene glycol) to Surfaces, Critical for Reducing Bacterial Adhesion. *Langmuir* **2003**, *19*, 6912–6921. [CrossRef]
7. Davies, D.G.; Chakrabarty, A.M.; Geesey, G.G. Exopolysaccharide Production in Biofilms - Substratum Activation of Alginate Gene-Expression by Pseudomonas-Aeruginosa. *Appl. Environ. Microb.* **1993**, *59*, 1181–1186. [CrossRef]
8. Davies, D.G.; Geesey, G.G. Regulation of the Alginate Biosynthesis Gene Algc in Pseudomonas-Aeruginosa during Biofilm Development in Continuous-Culture. *Appl. Environ. Microb.* **1995**, *61*, 860–867. [CrossRef]
9. Sutherland, I.W. Biofilm exopolysaccharides: A strong and sticky framework. *Microbiology* **2001**, *147*, 3–9. [CrossRef]
10. Stoodley, P.; Sauer, K.; Davies, D.G.; Costerton, J.W. Biofilms as Complex Differentiated Communities. *Annu. Rev. Microbiol.* **2002**, *56*, 187–209. [CrossRef]
11. Costerton, J.W.; Cheng, K.J.; Geesey, G.G.; Ladd, T.I.; Nickel, J.C.; Dasgupta, M.; Marrie, T.J. Bacterial Biofilms in Nature and Disease. *Annu. Rev. Microbiol.* **1987**, *41*, 435–464. [CrossRef]
12. Brown, M.R.W.; Gilbert, P. Sensitivity of biofilms to antimicrobial agents. *J. Appl. Bacteriol.* **1993**, *74*, 87S–97S. [CrossRef]
13. Vasickova, P.; Pavlik, I.; Verani, M.; Carducci, A. Issues Concerning Survival of Viruses on Surfaces. *Food Environ. Virol.* **2010**, *2*, 24–34. [CrossRef]
14. Nabi, G.; Wang, Y.; Hao, Y.; Khan, S.; Wu, Y.; Li, D. Massive use of disinfectants against COVID-19 poses potential risks to urban wildlife. *Environ. Res.* **2020**, *188*. [CrossRef]
15. Balmer, M.E.; Poiger, T.; Droz, C.; Romanin, K.; Bergqvist, P.A.; Muller, M.D.; Buser, H.R. Occurrence of methyl triclosan, a transformation product of the bactericide triclosan, in fish from various lakes in Switzerland. *Env. Sci. Technol.* **2004**, *38*, 390–395. [CrossRef]
16. Macherius, A.; Lapen, D.R.; Reemtsma, T.; Römbke, J.; Topp, E.; Coors, A. Triclocarban, triclosan and its transformation product methyl triclosan in native earthworm species four years after a commercial-scale biosolids application. *Sci. Total Environ.* **2014**, *472*, 235–238. [CrossRef] [PubMed]
17. Fang, J.-L.; Stingley, R.L.; Beland, F.A.; Harrouk, W.; Lumpkins, D.L.; Howard, P. Occurrence, Efficacy, Metabolism, and Toxicity of Triclosan. *J. Environ. Sci. Health Part C* **2010**, *28*, 147–171. [CrossRef]
18. Weatherly, L.M.; Gosse, J.A. Triclosan exposure, transformation, and human health effects. *J. Toxicol. Environ. Health Part B* **2017**, *20*, 447–469. [CrossRef]
19. Banerjee, I.; Pangule, R.C.; Kane, R.S. Antifouling Coatings: Recent Developments in the Design of Surfaces That Prevent Fouling by Proteins, Bacteria, and Marine Organisms. *Adv. Mater.* **2011**, *23*, 690–718. [CrossRef]
20. Kumar, A.; Vemula, P.K.; Ajayan, P.M.; John, G. Silver-nanoparticle-embedded antimicrobial paints based on vegetable oil. *Nat. Mater.* **2008**, *7*, 236–241. [CrossRef]
21. Li, W.-R.; Xie, X.-B.; Shi, Q.-S.; Zeng, H.-Y.; Ou-Yang, Y.-S.; Chen, Y.-B. Antibacterial activity and mechanism of silver nanoparticles on Escherichia coli. *Appl. Microbiol. Biotechnol.* **2009**, *85*, 1115–1122. [CrossRef]
22. Morones, J.R.; Elechiguerra, J.L.; Camacho, A.; Holt, K.; Kouri, J.B.; Ramírez, J.T.; Yacaman, M.J. The bactericidal effect of silver nanoparticles. *Nanotechnology* **2005**, *16*, 2346–2353. [CrossRef]
23. Sharma, V.K.; Yngard, R.A.; Lin, Y. Silver nanoparticles: Green synthesis and their antimicrobial activities. *Adv. Colloid Interface Sci.* **2009**, *145*, 83–96. [CrossRef] [PubMed]

24. Calabrese, G.; Petralia, S.; Franco, D.; Nocito, G.; Fabbi, C.; Forte, L.; Guglielmino, S.; Squarzoni, S.; Traina, F.; Conoci, S. A new Ag-nanostructured hydroxyapatite porous scaffold: Antibacterial effect and cytotoxicity study. *Mater. Sci. Eng. C* **2021**, *118*. [CrossRef]
25. Rai, M.K.; Deshmukh, S.D.; Ingle, A.P.; Gade, A.K. Silver nanoparticles: The powerful nanoweapon against multidrug-resistant bacteria. *J. Appl. Microbiol.* **2012**, *112*, 841–852. [CrossRef] [PubMed]
26. McShan, D.; Ray, P.C.; Yu, H. Molecular toxicity mechanism of nanosilver. *J. Food Drug Anal.* **2014**, *22*, 116–127. [CrossRef]
27. Feng, Q.L.; Wu, J.; Chen, G.Q.; Cui, F.Z.; Kim, T.N.; Kim, J.O. A mechanistic study of the antibacterial effect of silver ions on Escherichia coli and Staphylococcus aureus. *J. Biomed. Mater. Res.* **2000**, *52*, 662–668. [CrossRef]
28. Matsumura, Y.; Yoshikata, K.; Kunisaki, S.-I.; Tsuchido, T. Mode of Bactericidal Action of Silver Zeolite and Its Comparison with That of Silver Nitrate. *Appl. Environ. Microb.* **2003**, *69*, 4278–4281. [CrossRef] [PubMed]
29. Gupta, A.; Maynes, M.; Silver, S. Effects of halides on plasmid-mediated silver resistance in Escherichia coli. *Appl. Environ. Microbiol.* **1998**, *64*, 5042–5045. [CrossRef]
30. Sánchez-López, E.; Gomes, D.; Esteruelas, G.; Bonilla, L.; Lopez-Machado, A.L.; Galindo, R.; Cano, A.; Espina, M.; Ettcheto, M.; Camins, A.; et al. Metal-Based Nanoparticles as Antimicrobial Agents: An Overview. *Nanomaterials* **2020**, *10*, 292. [CrossRef]
31. Zhang, X.-F.; Liu, Z.-G.; Shen, W.; Gurunathan, S. Silver Nanoparticles: Synthesis, Characterization, Properties, Applications, and Therapeutic Approaches. *Int. J. Mol. Sci.* **2016**, *17*, 1534. [CrossRef]
32. Pulit-Prociak, J.; Banach, M. Silver nanoparticles–a material of the future . . . ? *Open Chem.* **2016**, *14*, 76–91. [CrossRef]
33. Piperno, A.; Mazzaglia, A.; Scala, A.; Pennisi, R.; Zagami, R.; Neri, G.; Torcasio, S.M.; Rosmini, C.; Mineo, P.G.; Potara, M.; et al. Casting Light on Intracellular Tracking of a New Functional Graphene-Based MicroRNA Delivery System by FLIM and Raman Imaging. *ACS Appl. Mater. Interfaces* **2019**, *11*, 46101–46111. [CrossRef] [PubMed]
34. Neri, G.; Scala, A.; Barreca, F.; Fazio, E.; Mineo, P.G.; Mazzaglia, A.; Grassi, G.; Piperno, A. Engineering of carbon based nanomaterials by ring-opening reactions of a reactive azlactone graphene platform. *Chem. Commun.* **2015**, *51*, 4846–4849. [CrossRef] [PubMed]
35. Neri, G.; Micale, N.; Scala, A.; Fazio, E.; Mazzaglia, A.; Mineo, P.G.; Montesi, M.; Panseri, S.; Tampieri, A.; Grassi, G.; et al. Silibinin-conjugated graphene nanoplatform: Synthesis, characterization and biological evaluation. *FlatChem* **2017**, *1*, 34–41. [CrossRef]
36. Cordaro, A.; Neri, G.; Sciortino, M.T.; Scala, A.; Piperno, A. Graphene-Based Strategies in Liquid Biopsy and in Viral Diseases Diagnosis. *Nanomaterials* **2020**, *10*, 1014. [CrossRef]
37. Hu, W.; Peng, C.; Luo, W.; Lv, M.; Li, X.; Li, D.; Huang, Q.; Fan, C. Graphene-Based Antibacterial Paper. *ACS Nano* **2010**, *4*, 4317–4323. [CrossRef]
38. Akhavan, O.; Ghaderi, E. Toxicity of Graphene and Graphene Oxide Nanowalls Against Bacteria. *ACS Nano* **2010**, *4*, 5731–5736. [CrossRef]
39. Liu, S.; Zeng, T.H.; Hofmann, M.; Burcombe, E.; Wei, J.; Jiang, R.; Kong, J.; Chen, Y. Antibacterial Activity of Graphite, Graphite Oxide, Graphene Oxide, and Reduced Graphene Oxide: Membrane and Oxidative Stress. *ACS Nano* **2011**, *5*, 6971–6980. [CrossRef]
40. Cobos, M.; De-La-Pinta, I.; Quindós, G.; Fernández, M.J.; Fernández, M.D. Graphene Oxide–Silver Nanoparticle Nanohybrids: Synthesis, Characterization, and Antimicrobial Properties. *Nanomaterials* **2020**, *10*, 376. [CrossRef]
41. Szaraniec, B.; Pielichowska, K.; Pac, E.; Menaszek, E. Multifunctional polymer coatings for titanium implants. *Mater. Sci. Eng. C* **2018**, *93*, 950–957. [CrossRef]
42. Tang, J.; Chen, Q.; Xu, L.; Zhang, S.; Feng, L.; Cheng, L.; Xu, H.; Liu, Z.; Peng, R. Graphene Oxide–Silver Nanocomposite As a Highly Effective Antibacterial Agent with Species-Specific Mechanisms. *ACS Appl. Mater. Interfaces* **2013**, *5*, 3867–3874. [CrossRef]
43. Zhao, R.; Kong, W.; Sun, M.; Yang, Y.; Liu, W.; Lv, M.; Song, S.; Wang, L.; Song, H.; Hao, R. Highly Stable Graphene-Based Nanocomposite (GO–PEI–Ag) with Broad-Spectrum, Long-Term Antimicrobial Activity and Antibiofilm Effects. *ACS Appl. Mater. Interfaces* **2018**, *10*, 17617–17629. [CrossRef] [PubMed]
44. Kellici, S.; Acord, J.; Vaughn, A.; Power, N.P.; Morgan, D.J.; Heil, T.; Facq, S.P.; Lampronti, G.I. Calixarene Assisted Rapid Synthesis of Silver-Graphene Nanocomposites with Enhanced Antibacterial Activity. *ACS Appl. Mater. Interfaces* **2016**, *8*, 19038–19046. [CrossRef]

45. Ocsoy, I.; Paret, M.L.; Ocsoy, M.A.; Kunwar, S.; Chen, T.; You, M.; Tan, W. Nanotechnology in Plant Disease Management: DNA-Directed Silver Nanoparticles on Graphene Oxide as an Antibacterial againstXanthomonas perforans. *ACS Nano* **2013**, *7*, 8972–8980. [CrossRef] [PubMed]
46. De Faria, A.F.; de Moraes, A.C.M.; Marcato, P.D.; Martinez, D.S.T.; Durán, N.; Filho, A.G.S.; Brandelli, A.; Alves, O.L. Eco-friendly decoration of graphene oxide with biogenic silver nanoparticles: Antibacterial and antibiofilm activity. *J. Nanopart. Res.* **2014**, *16*. [CrossRef]
47. Hummers, W.S.; Offeman, R.E. Preparation of Graphitic Oxide. *J. Am. Chem. Soc.* **1958**, *80*, 1339. [CrossRef]
48. Salavagione, H.J.; Goómez, M.N.A.; Martiínez, G. Polymeric Modification of Graphene through Esterification of Graphite Oxide and Poly(vinyl alcohol). *Macromolecules* **2009**, *42*, 6331–6334. [CrossRef]
49. Creighton, J.A.; Blatchford, C.G.; Albrecht, M.G. Plasma resonance enhancement of Raman scattering by pyridine adsorbed on silver or gold sol particles of size comparable to the excitation wavelength. *J. Chem. Soc. Faraday Trans. 2* **1979**, *75*, 790. [CrossRef]
50. Davidson, C.A.; Griffith, C.J.; Peters, A.C.; Fielding, L.M. Evaluation of two methods for monitoring surface cleanliness—ATP bioluminescence and traditional hygiene swabbing. *Luminescence* **1999**, *14*, 33–38. [CrossRef]
51. Barbera, L.; De Plano, L.M.; Franco, D.; Gattuso, G.; Guglielmino, S.P.P.; Lando, G.; Notti, A.; Parisi, M.F.; Pisagatti, I. Antiadhesive and antibacterial properties of pillar[5]arene-based multilayers. *Chem. Commun.* **2018**, *54*, 10203–10206. [CrossRef]
52. Nečas, D.; Klapetek, P. Gwyddion: An open-source software for SPM data analysis. *Open Phys.* **2012**, *10*. [CrossRef]
53. Yao, S.; Li, Y.; Zhou, Z.; Yan, H. Graphene oxide-assisted preparation of poly(vinyl alcohol)/carbon nanotube/reduced graphene oxide nanofibers with high carbon content by electrospinning technology. *RSC Adv.* **2015**, *5*, 91878–91887. [CrossRef]
54. Agnihotri, S.; Mukherji, S.; Mukherji, S. Size-controlled silver nanoparticles synthesized over the range 5–100 nm using the same protocol and their antibacterial efficacy. *RSC Adv.* **2014**, *4*, 3974–3983. [CrossRef]
55. Lin, J.-J.; Lin, W.-C.; Dong, R.-X.; Hsu, S.-H. The cellular responses and antibacterial activities of silver nanoparticles stabilized by different polymers. *Nanotechnology* **2012**, *23*. [CrossRef] [PubMed]
56. Iqbal, M.; Zafar, H.; Mahmood, A.; Niazi, M.B.K.; Aslam, M.W. Starch-Capped Silver Nanoparticles Impregnated into Propylamine-Substituted PVA Films with Improved Antibacterial and Mechanical Properties for Wound-Bandage Applications. *Polymers* **2020**, *12*, 2112. [CrossRef] [PubMed]
57. Mineo, P.; Abbadessa, A.; Mazzaglia, A.; Gulino, A.; Villari, V.; Micali, N.; Millesi, S.; Satriano, C.; Scamporrino, E. Gold nanoparticles functionalized with PEGylate uncharged porphyrins. *Dye. Pigment.* **2017**, *141*, 225–234. [CrossRef]
58. Neri, G.; Fazio, E.; Mineo, P.G.; Scala, A.; Piperno, A. SERS Sensing Properties of New Graphene/Gold Nanocomposite. *Nanomaterials* **2019**, *9*, 1236. [CrossRef]
59. Ajitha, B.; Kumar Reddy, Y.A.; Reddy, P.S.; Jeon, H.-J.; Ahn, C.W. Role of capping agents in controlling silver nanoparticles size, antibacterial activity and potential application as optical hydrogen peroxide sensor. *RSC Adv.* **2016**, *6*, 36171–36179. [CrossRef]
60. Saha, S.K.; Chowdhury, P.; Saini, P.; Babu, S.P.S. Ultrasound assisted green synthesis of poly(vinyl alcohol) capped silver nanoparticles for the study of its antifilarial efficacy. *Appl. Surf. Sci.* **2014**, *288*, 625–632. [CrossRef]
61. Ananth, A.N.; Daniel, S.C.G.K.; Sironmani, T.A.; Umapathi, S. PVA and BSA stabilized silver nanoparticles based surface–enhanced plasmon resonance probes for protein detection. *Colloids Surf. B Biointerfaces* **2011**, *85*, 138–144. [CrossRef]
62. Filippo, E.; Serra, A.; Manno, D. Poly(vinyl alcohol) capped silver nanoparticles as localized surface plasmon resonance-based hydrogen peroxide sensor. *Sens. Actuators B Chem.* **2009**, *138*, 625–630. [CrossRef]
63. Caccamo, D.; Currò, M.; Ientile, R.; Verderio, E.A.M.; Scala, A.; Mazzaglia, A.; Pennisi, R.; Musarra-Pizzo, M.; Zagami, R.; Neri, G.; et al. Intracellular Fate and Impact on Gene Expression of Doxorubicin/ Cyclodextrin-Graphene Nanomaterials at Sub-Toxic Concentration. *Int. J. Mol. Sci.* **2020**, *21*, 4891. [CrossRef] [PubMed]
64. Choudhary, P.; Parandhaman, T.; Ramalingam, B.; Duraipandy, N.; Kiran, M.S.; Das, S.K. Fabrication of Nontoxic Reduced Graphene Oxide Protein Nanoframework as Sustained Antimicrobial Coating for Biomedical Application. *ACS Appl. Mater. Interfaces* **2017**, *9*, 38255–38269. [CrossRef] [PubMed]

65. Kumara, C.; Luo, H.; Leonard, D.N.; Meyer, H.M.; Qu, J. Organic-Modified Silver Nanoparticles as Lubricant Additives. *Acs Appl. Mater. Interfaces* **2017**, *9*, 37227–37237. [CrossRef] [PubMed]
66. Kyrychenko, A.; Pasko, D.A.; Kalugin, O.N. Poly(vinyl alcohol) as a water protecting agent for silver nanoparticles: The role of polymer size and structure. *Phys. Chem. Chem. Phys.* **2017**, *19*, 8742–8756. [CrossRef]
67. Chang, A.; Schnall, A.H.; Law, R.; Bronstein, A.C.; Marraffa, J.M.; Spiller, H.A.; Hays, H.L.; Funk, A.R.; Mercurio-Zappala, M.; Calello, D.P.; et al. Cleaning and Disinfectant Chemical Exposures and Temporal Associations with COVID-19—National Poison Data System, United States, 1 January 2020–31 March 2020. *Mmwr. Morb. Mortal. Wkly. Rep.* **2020**, *69*, 496–498. [CrossRef]

Publisher's Note: MDPI stays neutral with regard to jurisdictional claims in published maps and institutional affiliations.

© 2020 by the authors. Licensee MDPI, Basel, Switzerland. This article is an open access article distributed under the terms and conditions of the Creative Commons Attribution (CC BY) license (http://creativecommons.org/licenses/by/4.0/).

Review

Graphene-Based Strategies in Liquid Biopsy and in Viral Diseases Diagnosis

Annalaura Cordaro [1], Giulia Neri [1], Maria Teresa Sciortino [1], Angela Scala [1,2] and Anna Piperno [1,2,*]

[1] Department of Chemical, Biological, Pharmaceutical and Environmental Sciences, University of Messina, 98166 Messina, Italy; acordaro@unime.it (A.C.); nerig@unime.it (G.N.); mtsciortino@unime.it (M.T.S.); ascala@unime.it (A.S.)

[2] Consorzio Interuniversitario Nazionale di ricerca in Metodologie e Processi Innovativi di Sintesi (C.I.N.M.P.I.S.), Unità Operativa dell'Università di Messina, 98166 Messina, Italy

* Correspondence: apiperno@unime.it; Tel.: +39-090-6765-173

Received: 25 April 2020; Accepted: 21 May 2020; Published: 26 May 2020

Abstract: Graphene-based materials are intriguing nanomaterials with applications ranging from nanotechnology-related devices to drug delivery systems and biosensing. Multifunctional graphene platforms were proposed for the detection of several typical biomarkers (i.e., circulating tumor cells, exosomes, circulating nucleic acids, etc.) in liquid biopsy, and numerous methods, including optical, electrochemical, surface-enhanced Raman scattering (SERS), etc., have been developed for their detection. Due to the massive advancements in biology, material chemistry, and analytical technology, it is necessary to review the progress in this field from both medical and chemical sides. Liquid biopsy is considered a revolutionary technique that is opening unexpected perspectives in the early diagnosis and, in therapy monitoring, severe diseases, including cancer, metabolic syndrome, autoimmune, and neurodegenerative disorders. Although nanotechnology based on graphene has been poorly applied for the rapid diagnosis of viral diseases, the extraordinary properties of graphene (i.e., high electronic conductivity, large specific area, and surface functionalization) can be also exploited for the diagnosis of emerging viral diseases, such as the coronavirus disease 2019 (COVID-19). This review aimed to provide a comprehensive and in-depth summarization of the contribution of graphene-based nanomaterials in liquid biopsy, discussing the remaining challenges and the future trend; moreover, the paper gave the first look at the potentiality of graphene in COVID-19 diagnosis.

Keywords: graphene; SERS; liquid biopsy; circulating tumor cells; exosomes; circulating nucleic acids; COVID-19

1. Introduction

Liquid biopsy is a minimally invasive technology for the detection of molecular biomarkers in blood and other body fluids (urine, saliva, ascites fluids, pleural effusions, etc.). The term was coined several decades ago, when was discovered, for the first time, the presence of extracellular nucleic acids in humans [1]; currently, it comprises not only the detection of extracellular/cell-free nucleic acids (NAs) with diagnostic significance but also of circulating tumor cells (CTCs) and extracellular vesicles (EVs), mainly exosomes (EXs). Although liquid biopsy cannot provide information related to tissue architecture and pathological microenvironment, it is considered a revolutionary technique that is opening unexpected perspectives in the early diagnosis and, in the therapy monitoring, severe diseases, ranging from cancer [2], metabolic syndrome [3], autoimmune disease [4], neurodegenerative disorders [5], and atherothrombosis [5] to prenatal screening [6].

Despite the high potential of liquid biopsy, the isolation, characterization, and quantification of NA, CTC, and EX biomarkers, due to their specific intrinsic features and low concentrations in the complex biological matrix, require complex procedures, and the systematic application in real practice is still hindered by many hurdles, such as unsatisfactory specificity and sensitivity, lack of standardization methods, and cost-effectiveness. Recently, a series of technological advancements in liquid biopsy has been obtained from the rapid development of nanotechnology-based strategies that provide a remarkable control over nanoparticle design, allowing to tailor their properties toward specific applications [7]. A plethora of nanomaterials, nanostructures, and molecular probes have been proposed for the fabrication of devices able to provide readable signals for early diagnosis and dynamic monitoring of diseases, taking advantage of their outstanding electrical, magnetic, optical, mechanical, or thermal characteristics at nanoscale dimensions [8]. Due to their unique physicochemical properties, arising from their high surface area, size, shape, unique optical properties, and surface chemistry, graphene-based materials (G) can realistically devise more advanced applications for liquid biopsy scope. The current review dealt with the recent advancements of G platforms for effectively capturing, identification, and quantification of NA, CT, and EX biomarkers. We discussed the main design criteria that have been used to develop multifunctional G platforms, bringing out the specific role of G in the selective capture and identification of heterogeneous biomarkers from the body fluids. Particular attention was reserved for the advances of liquid biopsy in cancer diagnosis and monitoring. Final remarks were devoted to challenges and the opportunity to adapt G technology for the diagnosis of emerging viral diseases, such as coronavirus disease 2019 (COVID-19).

2. Graphene Nanomaterials

The outcomes of graphene-based platforms in sensing applications are strictly correlated to the physicochemical properties of the starting material used for their fabrication [9]. However, a univocal classification of G broad family and their correlation with biosensing properties are challenging. Thus, the different synthetic approaches have been adopted for G preparation; the not homogeneous G nature (complexes structures with several oxidation states, varied lateral sizes, different number of layers, and different colloidal stability); the presence of impurities (often metal impurities); the formation of nanocomposites by a combination of G with organic or inorganic compounds have been taken in consideration for graphene-based biosensing applications [10].

G has been obtained by bottom-up or top-down approaches, differing for (i) the number and dimension of layers; (ii) the amount of oxygen functional groups scattered over the carbon surfaces; (iii) chemical features of compounds introduced during the post-synthetic decoration process, etc. [11–14].

Commonly, high-quality mono or multilayer G systems have been obtained by bottom-up approaches, such as epitaxial growth (EG) or chemical vapor deposition (CVD) on metallic substrates. These materials are endowed with ideal features (i.e., large surface area and high homogeneity) to be used as components of electronic devices. However, the high cost of these strategies, together with the requirement to transfer G on more suitable substrates, has limited the graphene's scale-up production [15].

Top-down strategies, such as chemical or physical exfoliation of graphite bulk, are regarded as valuable synthetic options to develop G for diagnostic devices [16,17]. G commonly used in the biosensing field includes graphene oxide (GO), reduced graphene (G-red), functionalized graphene (f-G), together with emerging derivatives, such as graphene quantum dot (GQD), N-doped multiple graphene aerogel, graphene field-effect transistor (GFET) etc. The plethora of G is continuously supplied by new derivatives with unique properties, which potentially enable an entirely new generation of technologies beyond the limits of conventional materials [18–20].

GO is obtained by chemical oxidation of graphite and successive exfoliation of graphite oxide via ultrasonication. Oxygen functionality groups on GO surfaces are widely exploited in the chemical functionalization of GO, especially by esterification/amidation reactions at the carboxylic groups [21–23]. Processability and water stability due to ionizable groups on GO surfaces are the main advantages in the use of GO; whereas the structural defects on the sp^2 network and the lacking electrical conductivity are the main limits for GO applications as an electronic device [24].

G-red is obtained from GO nanosheets by different techniques, including the solvothermal process or chemical reduction with hydrazine [25,26]. A partial restoring of the sp^2 network, which results in an improved electrical conductivity and mechanical strength of G-red, compared to GO, has been obtained by the reduction process. Nowadays, G-red stable colloidal systems are obtained by using biocompatible reducing agents, such as gallic acid, starch, vitamin C, etc., allowing to reduce the cost and the environmental impact [27].

GQDs are fluorescent carbon nanosystems, generally arising from G or GO, composed of less than ten graphene layers with a later dimension less than 10 nm. GQDs do not possess only the intrinsic properties of graphene but also new properties due to edge effects and significant quantum confinement [28]. A wide variety of GQDs is obtained by bottom-up or top-down approaches. In the first case, the adopted strategies are characterized by a good size control and by the possibility to tune the GQDs properties on the basis of substrate nature. However, they suffer from some drawbacks, i.e., the employment of toxic solvent, high temperature, and substrate concentrations. Top-down approaches give a large scale production of GQDs due to the early synthetic steps and the use of cheap carbon starting materials [29].

GQDs have shown lower toxicity and higher photostability compared to other semiconducting quantum dots, and several applications, ranging from catalyst to nanomedicine, have been proposed. In particular, electrochemical, optical, and photoelectrochemical biosensors based on GQDs, characterized by a high sensing selectivity, have been developed [30].

An emerging class of 3D carbon materials (aerogel, foam, hydrogel, etc.) have been recently proposed for water decontamination and as conversion/storage energy devices [31]. Template-assisted methods, based on CVD strategy or graphene/GO layers assembling processes, such as self-assembly of G-red sheets reduced via the solvothermal or hydrothermal method, have provided 3D graphene-based materials, characterized by the intrinsic properties of G together with new interesting physicochemical properties, such as high porosity, low density, unique electrochemical performance [32]. N-doping strategies have been widely adopted to tune the electrochemical properties of G derivatives. N-doped G has shown high performance like photocatalytic systems for the reduction of CO_2 and the degradation of organic contaminants under visible light [33].

The replacement of the traditional semiconductors-based electronic devices with a single layer graphene-based material has been proposed and used for the fabrication of GFET, proposed as sensors in physical, chemical, and biological application fields [34].

3. Tumor Biomarkers in Liquid Biopsy

Considering the temporal and spatial heterogeneity and its evolution, the tumor needs to be monitored at distinct times of the disease for an efficient treatment. Therefore, there is an urgent need to search for minimally invasive approaches in order to detect and monitor the disease progression throughout the treatment. Indeed, surgical tissue biopsies are invasive procedures, often difficult to perform on organs that lie deep within the body, and their use is limited as they can give false-negative results due to sampling. Therefore, it is necessary to identify ideal biomarkers that can be used for the early diagnosis, detection of recurrence, and monitoring of metastasis for cancer. A liquid biopsy might be a promising approach because it deals with the communication in tumor microenvironment. According to several research studies, the liquid biopsy is defined as the capture and the analysis of tumor-related biomarkers in a fluid sample. The biomarkers are represented by circulant tumor cells (CTCs), circulant tumor nucleic acids (ctNAs), proteins, and/or tumor-derived extracellular vesicles (EVs), which have been shed from tumor masses (Figure 1) into the bloodstream, saliva, urine, cerebrospinal fluid (CSF), among other peripheral fluids of patients. The liquid biopsy provides a more comprehensive snapshot of intra-tumor clonal heterogeneity compared to single-site tissue biopsies and, in addition, can allow repeated blood sampling, thereby providing an insight into the evolutionary dynamics of cancer. For these reasons, liquid biopsy should be extensively studied due to its minimal invasiveness and can be used for the early diagnosis and monitoring of metastasis in cancer patients [35].

The main approaches to liquid biopsies have embraced the detection of CTCs [36,37], the capture of exosomes (EXs) that are secreted by tumor mass [38], and the analysis of ctDNA or miRNA in body fluid samples [39] since the first studies. Indeed, due to the rapid turnover of cancer cells and the constant release of tumor-derived nucleic acids, vesicles, and viable CTCs into the circulation, the ability to detect and characterize has enabled surgeons to analyze the evolution of the tumor at distinct times and, most importantly, in a non-invasive manner. Literature data have demonstrated that levels of these biomarkers increase in patients with several malignant types of tumors, such as breast, ovarian cancer, stomach, colorectal, prostate, lung, and others. However, most studies have been done in patients with late-stage cancer, mainly due to the considerably higher concentrations of the above-mentioned biomarkers in their blood. Based on these promising findings, data from Wroclawski and collaborators demonstrated that serum DNA levels were significantly increased in patients with colorectal cancer of stage IV and fluctuated during chemotherapy [40]. Lung cancer patients, if compared to the control patients, have demonstrated significant differences in ctDNA levels since stage I [41]. The fluctuations of ctDNA were proposed by Diehl et al. as a biomarker to monitor the course of therapy in patients with metastatic colorectal cancer (mCRC) undergoing surgery or chemotherapy treatments [42]. The level of ctDNA has been quantified by BEAMing (beads, emulsions, amplification, and magnetics) and compared to carcinoembryonic antigen marker (CEA), routinely used in the management of the disease in subjects with CRC [42]. Numerous gastrointestinal diseases can also lead to an increase in ctDNA, even if considered malignant or benign.

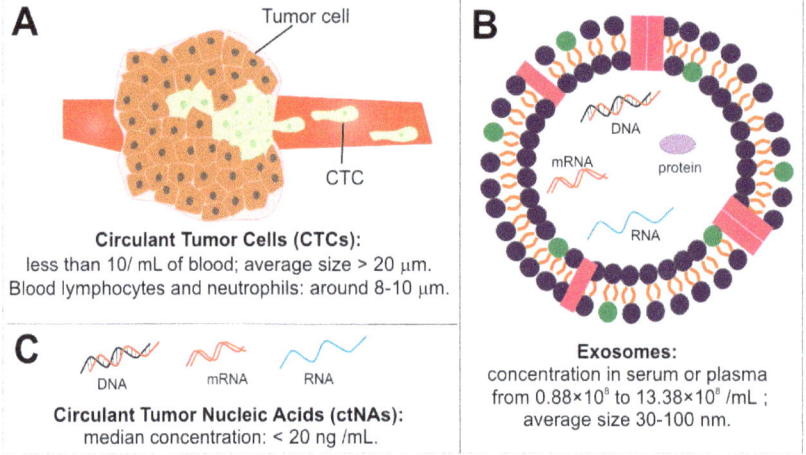

Figure 1. Schematic representation of typical cancer biomarkers of liquid biopsy: (**A**) Circulating tumor cells (CTCs); (**B**) Exosomes (EXs); (**C**) Circulant nucleic acids (ctNAs).

The diagnosis of ovarian cancer (OC) is mainly based on levels of biomarker CA-125 in blood and imaging. Recent data have shown that EVs possess advantages in terms of abundance, stability, and accessibility, compared with CTCs and ctDNA. Furthermore, the contents of EVs are tumor-specific and reveal a high correlation with tumor staging and prognosis [43]. Additionally, due to tumor heterogeneity, a panel of biomarkers will be more useful and reliable, instead of a single marker, for OC early diagnosis and screening high-risk individuals [44,45].

4. Circulant Tumor Cells (CTCs)

CTCs are a population of rare cancer cells detached from the primary tumor and shed into the bloodstream, becoming the main responsible for metastases in different organs. They are emerging as potential biomarkers and non-invasive alternative to tissue biopsy for the early detection, diagnosis, and prognosis of cancer, to improve the clinical settings of patients [46]. Since CTCs are extremely rare cells in the blood vessels (usually less than 10/mL of blood), their isolation from billions of red blood cells and millions of white blood cells and their accurate identification remain a challenge. Their heterogeneity (variety of surface protein expressions, sizes, and physical features), depending on the type and stage of cancer, makes laborious their isolation, requiring extremely sensitive and specific recognition methods [47]. In general, CTC detection includes four steps, such as capture, enrichment, detection, and the final release. The capture step is based on specific interactions between CTCs and materials (physical or antibody/antigen interaction). The enrichment step refers to CTCs isolation from the blood. The CTCs detection is generally carried out by fluorescence, surface-enhanced Raman scattering (SERS), or electrical impedance measurements. Finally, the enriched CTCs are released for further phenotype identification and molecular analyses [48]. To date, several technologies have been refined for CTCs detection, enrichment, and isolation, based on chemical or physical methods, such as capture by magnetic nanoparticles (NPs) [49], mechanical separation by size difference [50], microfluidic approaches [51,52], and immune-recognition methods [47].

Among the antibody-dependent isolation procedure based on specific biomarkers recognition, immunomagnetic technologies are often performed using anti-EpCAM antibody-functionalized magnetic NPs to specifically target EpCAM (Epithelial cell adhesion molecule) expressing cells. The epithelial cell adhesion molecule (EpCAM) is a transmembrane glycoprotein that mediates cell–cell adhesion in epithelial tissues. Since it has oncogenic potential, it has been extensively used for CTC capturing. To date, the only FDA approved CTC detection system is the CellSearch® assay, although the high-cost fabrication limits its use. The kit is based on immunomagnetic separation, to target a specific antigen by using anti-EpCAM antibodies coupled to magnetic beads. The subsequent separation of the antigen-antibody complex can be achieved via exposure to a magnetic field [53].

The development of reliable, cost-effective, and sensitive CTC detection and isolation technologies plays a pivotal role in the early diagnosis and treatment of cancer. Nanomaterials offer excellent advantages to improve the sensitivity in biomolecule detection due to their high surface area to volume ratio and similar size with respect to biomolecules [54]. Many classes of nanomaterials have been incorporated into CTC research for highly sensitive and selective cell capture, i.e., magnetic and gold nanoparticles, carbon nanotubes, dendrimers, quantum dots, and graphene oxide (GO) [55]. Specifically, recent progresses in nanoscience have allowed designing nanoarchitectures based on multifunctional G platforms for the isolation and identification of CTCs, representing technological advancements in liquid biopsy [56]. In fact, the ease surface chemical modifications, together with its unique optical properties, make GO an attractive material for biomolecule detection [56], and the most commonly used strategies to isolate CTCs are based on traditional immunomagnetic separation, electrochemical technology, and microfluidic tools [47]. Because of the diamagnetic feature of all untreated biological materials, magnetic cell separation using bio-conjugated magnetic materials can be fruitfully applied to separate CTCs from whole blood, in a highly specific way, via targeted binding and subsequent separation using a bar magnet, avoiding light scattering and autofluorescence background from blood cells.

The combination of graphene oxide quantum dots (GOQDs) and magnetic nanoplatforms into a single nanoarchitecture functionalized with anti-GPC3 (Glypican 3) antibody has been proposed for the accurate identification and selective capture of liver cancer tumor CTCs [57]. An electrochemical sensing strategy based on aptamer-functionalized and gold nanoparticles array-decorated magnetic graphene nanosheet (AuNPs-Fe$_3$O$_4$-GS) has been reported for monitoring and capturing CTCs in human whole blood. The sensor takes advantage of the combination of two effects: the efficient recognition and capture of the target CTCs assured by selected aptamers and the signal amplification guaranteed by the functionalization of the gold nanoparticles (AuNPs) with electroactive species (6-ferrocenyl-1-hexanethiol or thionine) [58].

Several GO-based microfluidic devices have been proposed to enrich CTCs, based on their distinct biochemical properties toward other human blood components. Most of these devices focus on immunoaffinity-based technologies, which exploit specific antibodies widely expressed in cancer cells to isolate CTCs with high purity and sensitivity.

A microfluidic GO-based chip with accurate surface capture design has been reported for isolating CTCs from metastatic breast cancer patients, with high sensitivity and reproducibility. The use of GO as the base material for antibody conjugation enables the chip to detect CTCs from only 1 mL of blood, with high yield and reproducibility due to the high-density antibody presentation [59].

A microfluidic device exploiting immunocapture based on a tunable thermal-sensitive polymer-GO chip has been proposed for highly efficient capture and subsequent release of CTCs from breast and pancreatic cancer patients. The microfluidic device is coated with a composite film of functionalized GO dispersed in a thermoresponsive polymer matrix. The combination of a biocompatible GO, properly functionalized for immunocapture, with a thermosensitive polymer, has provided temperature-dependent modulation of capture/release, allowing the effective cell release for post-capture analysis. This device has overcome the common drawback of most immunoaffinity-based technologies reliant on antibodies attached to the capture surface, hindering the release of viable cells [60]. Electrochemical technology is also applicable to CTC recovery. A graphene-based electrochemical sensing platform, based on functionalized graphene-modified glassy carbon electrodes (GCEs), has been designed to be incubated with mammalian cells (i.e., different cancerous, multidrug-resistant cancerous, and metastatic human breast cells, as well as artificial CTC samples). The interactions with cell surface components, responsible for conjugating the target cells on the electrode surface, have been transduced to an ultrasensitive electrochemical response. The chemical diversity offered by the graphene probes has allowed discerning different cell surface/cell type, serving as a sensor array featuring selective receptors. The advantage of such an array-based sensing approach relies on the possibility to make an overall signature of CTCs, providing a fingerprint that allows for the classification and identification of cells [61]. A porous graphene-oxide (PGO) has been used to decorate light addressable-potentiometric-sensor (LAPS) surface, followed by the aptamer AS1411 anchoring (apta-PGO-LAPS), and is investigated as a light addressable potentiometric sensor. The CTC sensing interface has exploited the integration of electronic sensors with the robust and specific CTC's bioprobe (aptamer). Specifically, the aptamer probe AS1411 has owned high binding affinity and specificity to the overexpressed nucleolin on the CTCs' membrane [62].

A sensor, for clinical sample's CTC detection, based on aptamer AS1411 functionalized graphene field-effect transistor (GFET) by using tetra (4-aminophenyl) porphyrin-mediated reduced GO as the channel material, has been recently proposed. The aptamer sensing strategy has been applied to isolate CTCs of human lung cancer cell line A549, breast cancer MDAMB-231, and cervical cancer HeLa, with good sensitivity [63].

A versatile super-sandwich cytosensor, based on GO-modified 3D microchip and Au-enwrapped silica nanocomposites (Si/AuNPs), fabricated by photolithography, has been developed as CTC-sensitive quantitative detection system. The sensor integrates two functional components: (1) an anti-EpCAM coating on GO for recognizing/capturing EpCAM-expressing cells, and (2) horseradish peroxidase (HPR) and anti-CA15-3 (Ab2) loaded in Si/AuNPs to improve the selectivity of target cells and amplify the electrochemical detection signal. The performance was assessed on MCF7 breast cancer cells, showing high sensitivity with a wide range of 10^1 to 10^7 cells mL^{-1} and a detection limit of 10 cells mL^{-1} [64]. A CTC isolation platform based on GO functionalized polyester fabric sheets bearing anti-EpCAM antibodies has been proposed as a low-cost, easy-to-fit, and disposable platform, assuring high sensitivity. Capture efficiency of 75–80% was obtained for cells with high EpCAM expressions [65].

A 3D hierarchical nanostructured graphene cell-captured foam with an anti-EpCAM coating (rGO/ZnO/anti-EpCAM foam) has been proposed for recognizing/capturing EpCAM-expressing cancer cells, showing some advantages compared to microfluidic-based devices, such as easy fabrication, increased cell-substrate contact frequency in all directions, microporosity, which allows normal red blood cells to travel through, but selectively captures CTCs, due to the anti-EpCAM coating. The performance of this 3D foam was investigated using EpCAM-positive cancer cell lines (MCF7, breast cancer cells), resulting in a cell-capture yield reaching up to 58% after an incubation time of 30 min [66].

For more comprehensive CTC enrichment, special attention must be focused on the choice of antibodies. By combining different antibodies in a single nanodevice, higher capture efficiency can be achieved than that obtained by single biomarker recognition. Reduced graphene oxide (rGO) films functionalized with anti-EpCAM and anti-prostate specific membrane antigen (anti-PSMA) antibodies have been recently fabricated by spray coating rGO solution onto a smooth glass slide. The antibody-modified rGO films exhibited a high efficiency (60%) of CTC capture from the blood of prostate cancer patients with prostate-specific antigen (PSA) levels of 4–10 ng mL^{-1} [67].

5. Exosomes (EXs)

EXs are a subgroup of cell-derived nano-sized (30–100 nm) extracellular vesicles (EVs) that have been recently recognized as new mediators for many cellular processes and potential biomarkers for non-invasive disease diagnosis and for monitoring treatment response, especially in cancer therapy. Mounting evidences have demonstrated the EX implication in several diseases, including viral pathogenesis [68], neurodegenerative diseases [69], and cancer growth and progression [70]. In particular, the release of EXs has been found to increase significantly in most neoplastic cells and occurs continuously at all stages of tumor development. Growing evidence has shown that the tumor-derived EXs carry characteristic proteins and RNAs in various cancer types, and the expression levels of these molecules are closely correlated with tumor progression [71]. Besides, the surface protein expression can provide invaluable information associated with the physiological states of the parental cells, that is why EXs are emerging as a novel disease diagnosis tool. Although EXs share several protein markers on their membrane, some of them are cell-specific and reflect the conditions of the secreting cell, meaning that there is a large heterogeneity among these biological markers in a single sample of withdrawn blood; this makes their isolation rather difficult. Up to date, most of the microfluidic devices are still not compatible with clinical analysis due to scalability, standardization, and validation. Further, several approaches are time-consuming, require extensive pre-treatment steps, and do not recover enough samples for genomic or proteomic analysis. Thus, there is a need for isolation techniques that selectively isolate EXs in a cost-effective and rapid manner [2]. Nevertheless, the performance of common isolation methods is significantly affected by contamination from other membrane-derived subcellular structures with high similarity in physical properties, resulting in very poor recovery yields. Numerous EX isolation techniques have been established so far, including ultracentrifugation, polymer-based precipitation, filtration, and affinity pull-down. Currently, the most common method for EX purification is the ultracentrifugation, which includes several centrifugation steps. Polymer-based precipitation relies on the formation of a polymer network to entangle all lipid components in the sample and to reduce their solubility for rapid removal under a low centrifugal force [72–74]. Membrane filtration has also been applied for size-based isolation of EXs, but EXs are prone to adhere to the filtration membranes, causing sample loss. Moreover, the additional force applied to pass the analyzed liquid through the membranes could potentially deform or damage the EXs [74]. Affinity pull-down is superior in selective separation of EXs using specific antibodies, but it requires large amounts of sample volumes.

The development of "bio-sensors" able to recognize EXs, without purification steps, from biological samples with very high accuracy and sensitivity, has recently spread among the scientific community. Generally, they combine the specificity of immunoaffinity-based systems with functionalized nanomaterials. Fang et al. [71] designed a hybrid platform that integrated two nanomaterials with different surface properties: the hydrophilic macroporous graphene foam (GF) and the amphiphilic periodic mesoporous organosilica (PMO). The high specific surface area of GF, after modification with the antibody against the EX protein marker, CD63 (specific exosome marker), allowed highly specific isolation of EXs from complex biological samples with high recovery. After lysis with methanol, the amphiphilic PMO was employed to rapidly recover the EX proteins, including the highly hydrophobic membrane proteins. Peptides obtained by protein digestion were analyzed by LC-MS/MS analysis (liquid chromatography-tandem mass spectrometry). Zhang et al. reported a microfluidic platform based on the graphene oxide/polydopamine (GO/PDA) system [75]. GO induced spontaneous polymerization of a 3D PDA surface coating, which was demonstrated to improve the efficiency of EX immuno-capture, suppressing the effects of non-specific adsorption. The platform was prepared by a layer-by-layer coating method (Figure 2), and the on-chip-captured EXs were detected by fluorescence analyses after the treatment with a mixture of biotinylated antibodies (CD63, CD81, and EpCAM). Streptavidin-conjugated β-galactosidase (SβG) was used as a reporter enzyme. The platform performance was proved in both molecular profiling and the quantitative EXs detection of purified samples from a colon cancer cell line or directly in plasma samples from ovarian cancer patients. Surface-enhanced Raman scattering (SERS) spectroscopy is a promising analytic tool for EXs' ultra-detection. SERS' biomedical applications include two general methodologies, called label-free detection and indirect approaches based on the use of a Raman reporter (RaR) linked to nobel NPs, commonly known as SERS tags or SERS-labeled NPs [76]. A SERS tag consists of four main components: (1) silver or gold NPs, which act as plasmonic enhancer; (2) Raman reporter (RaR) acting as fingerprint label; (3) a protective layer or coating shell that stabilizes the NPs, allowing the biomolecules grafting; (4) recognition moieties. Noble metal NPs inducing an enhanced electric field when LSPR (localized surface plasmon resonance) is excited by selected laser light sources [16] might be considered the SERS tag core. The RaR, an organic compound with a typical spectral fingerprint (i.e., benzenthiol, 4-mercaptobenzoic acid, etc.), ideally should cover the NPs to provide a stable, intense, and reliable Raman signal. The coating component of SERS-tag, although not essential, can improve the colloid stability and provide several advantages: (a) prevent the RaR leaching; (b) avoid contaminations; (c) reduce the intensity variations due to NP-NP interactions. Several protective coatings have been proposed, including biomolecules (i.e., bovine serum albumin), polymers (i.e., PEG), inorganic shell (i.e., SiO_2), liposomes [76], and graphene [77]. In the last case, G acts as both a protective shell and RaR. Specific peptides, antibodies, or proteins are grafted in the external layer of the SERS-tag as recognition ligands of biomarkers [76].

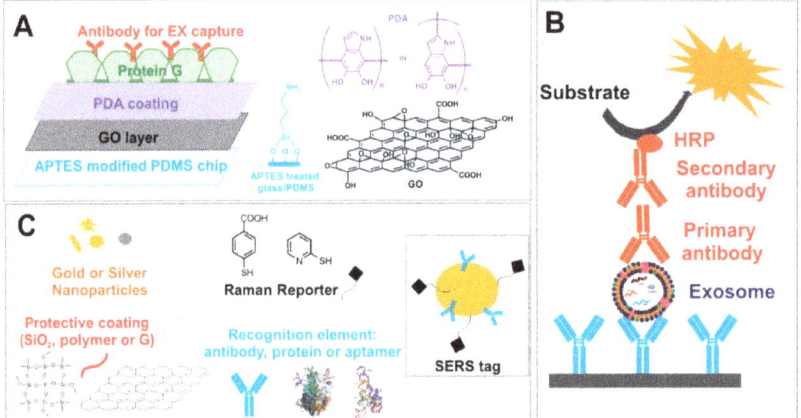

Figure 2. (**A**) Representative structure of the microfluidic platform based on graphene oxide/polydopamine (GO/PDA) for the immunocapture of exosome. (**B**) Representative EXs detection by ELISA colorimetric test using horseradish peroxidase (HRP) as a reporter enzyme in the secondary antibody. (**C**) Surface-enhanced Raman scattering (SERS) tag with related SERS tag toolbox.

6. Circulant Tumor Nucleic Acids (ctNAs)

Circulating tumor nucleic acids (ctNAs), such as circulating cell-free DNA, RNA, microRNA (Figure 1), represent an innovative tool for liquid biopsy applications [2,78,79]. The ctNA levels, compared to other circulating free biomarkers (such as CTCs), are detectable early in the bloodstream; therefore, they can be used for the initial tumor detection and the disease monitoring [80]. To date, several ctNA detection approaches based on fluorescence, SERS spectroscopy, radiochemical, enzymatic approach, chemiluminescence have been investigated [46,81,82]. Unfortunately, the detection of these biomarkers is challenging due to their small size and low concentrations in body fluids [80]. Different ultra-sensitive detection methods, including nucleic-acid sequence-based amplification [83], rolling circle amplification (RCA) [84], and polymerase chain reaction (PCR) [85], have been proposed. However, the complexity, the high cost of reagents and equipment, and the time-consuming protocols prevent the translation of these strategies in the market. To overcome some of these disadvantages, also the electrochemical approaches, characterized by a lower cost and a high detection sensitivity via signal amplification, have been proposed in the last years [86].

GO and rGO have been chosen as sensing platforms to detect circulating oligonucleotides and cells by fluorescent spectroscopy due to their ability to adsorb single-stranded (ss) oligonucleotides by noncovalent approaches (π-π interactions and/or hydrogen bonds) [87]. At the same time, GO and rGO have shown a lower affinity towards the double-stranded (ds) oligonucleotides due to poor accessibility of nucleobases inside the double helix and a lower ability to adsorb longer oligonucleotides due to lower diffusivity processes [88]. Moreover, G materials are able to almost completely quench the fluorescence emission of the fluorescent dye linked to ss oligonucleotide. In the presence of complementary target oligonucleotides (circulating oligonucleotides), the ss oligonucleotides marked with fluorescent dye can be released by G surface and complexed with the complementary target oligonucleotide, restoring the fluorescent emission, allowing the identification of the circulating oligonucleotide fragments.

A biosensing platform able to simultaneously detect and evaluate the amounts of miR-141 and miR-21 (two miRNAs overexpressed in the early and in the advanced stage of prostate cancer) from several body fluids (blood, urine, saliva) was investigated by Salih Hizir et al. [89]. The ability of GO to adsorb ss DNA on its surface and to quench fluorescence emission was exploited for the design of GO platform engineered with two fluorophore-labeled antisense DNA strands: fluorescein amidites (FAM)-labeled anti-miR-21 and Cy5-labeled anti-miR-141. The platform resulted in a fluorescent quenching at 520 nm (FAM channel) and 670 nm (Cy5 channel). In the presence of overexpressed miR-21, a fluorescent signal enhancement at 520 nm was observed, whereas overexpressed miR-141 induced a fluorescent signal increase at 670 nm. Non-target miRNAs induced only a lower fluorescent increase at these channels; therefore, the increase of fluorescent signal at 520 or 670 nm indicated the presence of miR-21 or miR-141, and the increase of intensity fluorescence signal level was used to determinate the concentration of miR-21 and miR-141 fragments. The system was proposed not only to detect prostate cancer disease but also to evaluate its advancement stage. Unfortunately, the low sensitivity and low specificity are typical problems of these nanodevices, hindering their clinical application [89].

A new GO-polymer-oligonucleotide (nGO-PEGMA/M2) and enzyme (*DNase I*) system able to detect miR-10b in an RNA pool taken from metastatic breast cancer cells were reported by Robertson et al. [90]. The insertion of a specific mismatch fragment into the probe sequence induced an increase of specificity towards miR-10b, an oligonucleotide overexpressed in breast cancer. The nGO-PEGMA/M2 *DNase I* system was able to distinguish miR-10b from miR-10a, which differed only for a single nucleotide. The presence of the endonuclease *DNase I* improved the fluorescent sensitivity of the probe but also the background fluorescent signal. To overcome this drawback, the edge of GO was functionalized with PEGMA, which hindered the access of *DNase I* on the GO surface to avoid the increase of fluorescence background signal due to undesired enzymatic activity [90].

The combination of the quenching properties of GO and cyclic enzymatic amplification method (CEAM) has allowed developing GO/ssDNA probes able to detect and discriminate among several mir-21 miRNAs in cell lysate media. The up-regulation expression of mir-21 miRNAs is involved in solid tumor growth. The biological media have been obtained from lung carcinoma cell line A-549 and mammary epithelial cells MCF-10A. The presence of complementary miRNA has induced the restoration of fluorescence due to miRNA/DNA complex formation, previously quenched by GO. Subsequently, miRNA released from the *DNase I* digestion can complex with another ssDNA probe on the GO surface to start another cycle, enhancing the fluorescent signal until all released ssDNA probes are completely consumed [91].

In the presence of divalent salt, GO is not able to discriminate between ssNAs and dsNAs [92]. On the contrary, rGO has shown a higher selectivity towards miRNA compared to GO in the same adsorption conditions [93]. Taking into account these findings, Yan et al. developed a magnetic system based on rGO (magnetic beads@APTES@rGO) able to selectively adsorb miRNA from the RNA pool isolated from healthy human plasma [88]. Magnetic beads were employed to obtain a faster extraction process by centrifugation. Moreover, in situ reverse transcriptions (RT), such as rolling circle amplification (RCA) strategy, were applied to desorb and detect miRNA by rGO surface [88].

Several challenges have been also focused on the detection of both circulating ss/ds DNA. Ruiyi et al. developed a nitrogen-doped multiple graphene aerogel/gold nanostar biosensor (N-doped MGA/GNS) able to detect dsDNA by human serum via electrochemical approach [94]. The hybrid N-doped MGA/GNS system showed an electrocatalytic activity towards Fe $(CN)_6^{3-/4-}$ improved in the presence of dsDNA, which was demonstrated by amperometric detection. The authors ascribed this behavior to the interaction between DNA and under-coordinated Au(I) sites bonded on the N-doped MGA-5 surface [94]. Another electrochemical biosensor composed of G decorated with Au nanorods and polythionine film (G/Au NR/PT) deposited onto glassy carbon electrode (GCE) was developed by Huang et al. for the detection of human papillomavirus (HPV) DNA in human serum [95]. G was used to enhance the surface area and the electric conductivity of the system; Au NRs (Au nanorods) were employed to increase the immobilization of DNA probe; polythionines were selected due to their good electron transfer ability and due to their ability to bond the Au NR surfaces by their amine groups. The thiolated capture probes (CP) were immobilized on the biosensor via electrostatic interactions and Au–S covalent bonds. CP was hybridized with one terminal of DNA target (TD), which arose from HPV-16 long terminal repeat sequences. Moreover, two auxiliary probes (AP) were developed to complex TD (fragment to be detected in human serum) by a long-range self-assembly process. Finally, the 1,10-phenanthrolineruthenium dichloride ($[Ru(phen)_3]^{2+}$) was used as an electrochemical indicator due to its ability to bond the DNA by electrostatic interactions. The increase of electrochemical response signal depended on the amount of ($[Ru(phen)_3]^{2+}$) bonded to DNA nanostructure. Worth noticing, the two AP sequences could bond with each other on the biosensor surface, giving rise to considerable lengthy self-assembled DNA nanostructure, only in the presence of TD [95].

7. Graphene-Based Strategies in the Diagnosis of Viral Diseases

Direct methods, exploiting graphene nanotechnology, for the rapid virus detection, have been only marginally investigated in the past, and no critical discussion has been reported in successive literature reviews [96,97].

This attitude was unchanging even during SARS–CoV-1 emergency that was responsible for the 2003 severe acute respiratory syndrome (SARS) infection in Asia, causing about 8000 cases and 774 deaths, also during the Middle East respiratory distress syndrome (MERS) of 2013, which affected Saudi Arabia causing close to 858 deaths [12,98–100]. Advances in nanotechnology have begun to play an important role in viral detection, to improve the detection limit, operational simplicity of viral diagnostics [78].

A coplanar-gate graphene field-effect transistors (GFETs) [71] have been proposed for the detection of HIV-1 (human immunodeficiency virus 1) and MLV (murine leukemia virus) viruses using antibodies of vesicular stomatitis Indiana virus (VSV) as biorecognition element. VSV antibodies are immobilized on the G layer using 1-pyrenebutanoic acid succinimidyl ester (PASE). PASE binds G by π-π interactions, anchoring the antibody's primary amine groups by the opposite succinimidyl group. The formation of the virus-antibody complex leads to a downward shift of the Dirac point voltage, regardless of the types of detected viruses. The proposed platform has worked in a wide range of concentrations (from 47.8 aM to 10.55 nM), but the lack of virus specificity appears the main limitation of this strategy.

An surface plasmon resonance (SPR) sensor based on an polyamidoamine-functionalized rGO(composite, with monoclonal antibodies immobilized on self-assembled dithiobis (succinimidyl undecanoate, DSU) for the detection/quantification of Dengue virus (DENV), has been recently described [97].

The specificity and the sensibility of the sensor have been achieved by anchoring a stable biorecognition element (antibodies (IgM) against Dengue type 2 envelope proteins) on the gold surface of the sensor. The specific binding of antibody-DENV 2 E-protein allows a significant change in the angle of the reflectivity minimum that is correlated to Dengue virus detection. The proposed sensor has shown a sensitive and selective response towards DENV 2 E-proteins compared to DENV 1 E-proteins and ZIKV (Zika virus) E-proteins. Although no G materials have been integrated into the above-described sensor [97], the criteria used for its fabrication were included in this review since the strategy could be extended to other viruses, and the performance of SPR noble metal could be improved in the presence of G [76].

Differently from the past, the current sanitary pandemic emergency caused by the new type of coronavirus (SARS–CoV-2) is characterized by global effort to identify biomarkers that predict the severity of COVID-19 patients and to develop diagnostic tools for the rapid detection of SARS–CoV-2 infection [101].

Currently, nucleic acid testing on respiratory specimens is the reference gold standard method for the diagnosis of COVID-19 infected patients [102]. The test requires a series of laboratory procedures: (a) viral RNA extraction; (b) addition to a master mix containing nuclease-free water, reverse primers, a fluorophore-quencher probe, and a reaction mix (i.e., polymerase, reverse transcriptase, magnesium, nucleotides, and additives); (c) loading of extracted RNA/master mix into a PCR thermocycler; (d) several cycles at settled temperature. During the RT-PCR cycles, the cleavage of the fluorophore-quencher probe generates a fluorescent signal detected and recorded in real-time [101].

RT-PCR uses respiratory samples to genetically detect SARS–CoV-2; some data have suggested that 20–34% of COVID-19 patients resulted negative in the test despite being infected. This variance in the sensitivity could be mainly attributed to low viral load (i.e., patients tested in the early stage of the viral disease) [102]. Other RT-PCR issues include the time consuming and expensive analysis and the technical expertise in carrying out the text.

Other technologies, such as point-of-care technologies and serologic immunoassays, are rapidly emerging to address these deficiencies [78].

Analytic methods to assess prior infection and immunity to SARS–CoV-2 by antibody identification are essential for epidemiologic studies, although sensibility and specificity of the tests currently available in the market remain undefined. Cross-reactivity of antibody to non–SARS–CoV-2 coronavirus proteins is the main issue of these serologic tests [101,102]. The development of an antigen detection test [102] could take advantage of progress in the production of monoclonal antibodies against the nucleocapsid protein of SARS–CoV-2. The global effort to increase SARS–CoV-2 testing capacity takes advantage of the most recent advances in chemistry, molecular biology, genome technology, and nanotechnology. Several projects are ongoing in this direction, and some results are already reported in the literature [95,103].

The detection of SARS–CoV-2 in respiratory samples has been achieved by LSPR biosensor, combining the photothermal effect and plasmonic sensing transduction for SARS–CoV-2 viral nucleic acid [103].

A field-effect transistor (FET)-based biosensing device for detecting SARS–CoV-2 spike protein (S) in clinical samples was reported by Seo et al. [104]. Antibodies against S protein were anchored to the graphene sheet (external coating of FET) by 1-pyrenebutanoic acid succinimidyl ester (PBASE, Figure 3).

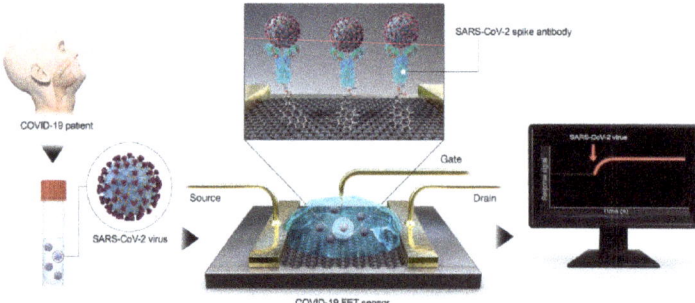

Figure 3. Coronavirus disease 2019 (COVID-19) field-effect transistor (FET)-sensor. Graphene is selected as sensing material and is decorated with the SARS–CoV-2 spike antibody using 1-pyrenebutanoic acid succinimidyl ester (PBASE) as interfacing molecule and probe linker. Reprinted with permission from reference [104], Copyright © 2020, American Chemical Society.

The performance of the sensor is determined using antigen protein, cultured virus, and nasopharyngeal swab specimens from COVID-19 patients. The device could detect S protein at concentrations of 1 fg/mL in PBS and 100 fg/mL in the clinical transport medium, and it could distinguish the SARS–CoV-2 antigen protein from those of MERS-CoV. The successful fabrication of a COVID-19 FET sensor based on the integration of the SARS–CoV-2 spike antibody with graphene suggests the key role of G for diagnostic scope [80]. Specifically, the functionalization of G with diverse functional molecules [14,17,105,106] could be the key element to tailor its properties and to obtain advanced diagnostic tools for the SARS–CoV-2 diagnosis. Meanwhile, for the revision of this manuscript, some works dealing with sensors for COVID-19 diagnosis based on graphene are reported in the literature [107], and, although further researches are undoubtedly necessary, the leading role of G in the world's fight against COVID-19 is clearly coming out [108].

In summary, the biomolecules till now used to target SARS–CoV-2 includes the viral RNA, the viral spike proteins, and the specific immunoglobulins produced by the host immune system. The biosensing community is actively working to improve portability, time, and cost of PCR-based SARS–CoV-2 detection, as well as to create manufacturable PCR-based microfluidic devices. Recently, also the gene-editing technology (CRISPR/Cas) has been developed to overcome the issues of PCR-based systems. Two different detection modes have been proposed in CRISPR technology, i.e., binding- or cleavage-based [109]. The sensor is developed by immobilization on a graphene-based field-effect transistor (GFET) of Cas9 with a sgRNA, specific to the target sequence of SARS COV-2; the electrical signal originated by the binding of the target nucleic acid by the Cas9–sgRNA complex is recorded via a simple handheld device without amplification.

8. Conclusions

The extraordinary properties of G make it a potential candidate to be routinely implemented in the design of biosensing platforms for liquid biopsy. Certainly, the innovation in diagnosis and monitoring of severe diseases could take advantage of the most recent progress in chemistry, molecular biology, genome technology, and nanotechnology. However, to give a significant contribution to the topics of great relevance for public health, such as cancer-fighting, neurodegenerative pathologies, emerging viral diseases, etc., the priority of collaborative research should be mainly focused on the opportunity to clinically translate the newly identified biomarkers using nanotechnology. Significant advancement has been achieved in the last years; however, data reproducibility remains the main drawback, and the selection of suitable nanomaterials for the development of devices is one of the key elements to obtain diagnostic tools that guarantee reproducible and reliable quantitative measurements.

Finally, the COVID-19 lesson indicates that the development of diagnostics is crucial to managing the pandemic outbreak, and certainly, G technology will assume a prominent role in the fabrication of innovative devices for the detection/quantification of viral nucleic acids/proteins actionable for detection at the point-of-care.

Author Contributions: Conceptualization and methodology A.C., G.N., M.T.S., A.S., A.P.; formal analysis, A.C., G.N., M.T.S., A.S., A.P.; resources, A.C., G.N., M.T.S., A.S., A.P.; data curation, A.C., G.N., M.T.S., A.S., A.P.; writing—original draft preparation, A.P.; writing—review and editing, A.C., G.N., M.T.S., A.S., A.P.; supervision, A.P., project administration, A.P.; funding acquisition, A.P. All authors have read and agreed to the published version of the manuscript.

Funding: This work was partially aided by PON Project *"Biopsie liquide per la Gestione Clinica dei Tumori"* (BiLiGeCT, ARS01-00492).

Conflicts of Interest: The authors declare no conflict of interest.

References

1. Schmidt, B.; Fleischhacker, M. Is liquid biopsy ready for the litmus test and what has been achieved so far to deal with pre-analytical issues? *Transl. Cancer Res.* **2018**, *7*, S130–S139. [CrossRef]
2. Vaidyanathan, R.; Soon, R.H.; Zhang, P.; Jiang, K.; Lim, C.T. Cancer diagnosis: From tumor to liquid biopsy and beyond. *Lab Chip* **2019**, *19*, 11–34. [CrossRef]
3. La Marca, V.; Fierabracci, A. Insights into the diagnostic potential of extracellular vesicles and their mirna signature from liquid biopsy as early biomarkers of diabetic micro/macrovascular complications. *Int. J. Mol. Sci.* **2017**, *18*, 1974. [CrossRef]
4. Stefancu, A.; Badarinza, M.; Moisoiu, V.; Iancu, S.D.; Serban, O.; Leopold, N.; Fodor, D. Sers-based liquid biopsy of saliva and serum from patients with Sjögren's syndrome. *Anal. Bioanal. Chem.* **2019**, *411*, 5877–5883. [CrossRef]
5. Suades, R.; Padró, T.; Crespo, J.; Sionis, A.; Alonso, R.; Mata, P.; Badimon, L. Liquid biopsy of extracellular microvesicles predicts future major ischemic events in genetically characterized familial hypercholesterolemia patients. *Arterioscler. Thromb. Vasc. Biol.* **2019**, *39*, 1172–1181. [CrossRef]
6. Macías, M.; Alegre, E.; Díaz-Lagares, A.; Patiño, A.; Pérez-Gracia, J.L.; Sanmamed, M.; López-López, R.; Varo, N.; González, A. Chapter three-liquid biopsy: From basic research to clinical practice. In *Advances in Clinical Chemistry*; Makowski, G.S., Ed.; Elsevier: Amsterdam, The Netherlands, 2018; Volume 83, pp. 73–119.
7. Li, W.; Ye, L.; Li, S.; Yao, H.; Ade, H.; Hou, J. A high-efficiency organic solar cell enabled by the strong intramolecular electron push–pull effect of the nonfullerene acceptor. *Adv. Mater.* **2018**, *30*, 1707170. [CrossRef]
8. Gribko, A.; Künzel, J.; Wünsch, D.; Lu, Q.; Nagel, S.M.; Knauer, S.K.; Stauber, R.H.; Ding, G.-B. Is small smarter? Nanomaterial-based detection and elimination of circulating tumor cells: Current knowledge and perspectives. *Int. J. Nanomed.* **2019**, *14*, 4187–4209. [CrossRef]
9. Zhu, Y.; Murali, S.; Cai, W.; Li, X.; Suk, J.W.; Potts, J.R.; Ruoff, R.S. Graphene and graphene oxide: Synthesis, properties, and applications. *Adv. Mater.* **2010**, *22*, 3906–3924. [CrossRef]
10. Gu, H.; Tang, H.; Xiong, P.; Zhou, Z. Biomarkers-based biosensing and bioimaging with graphene for cancer diagnosis. *Nanomaterials* **2019**, *9*, 130. [CrossRef]
11. Criado, A.; Melchionna, M.; Marchesan, S.; Prato, M. The covalent functionalization of graphene on substrates. *Angew. Chem. Int. Ed.* **2015**, *54*, 10734–10750. [CrossRef]
12. Lee, J.-H.; Park, S.-J.; Choi, J.-W. Electrical property of graphene and its application to electrochemical biosensing. *Nanomaterials* **2019**, *9*, 297. [CrossRef] [PubMed]
13. Neri, G.; Scala, A.; Barreca, F.; Fazio, E.; Mineo, P.G.; Mazzaglia, A.; Grassi, G.; Piperno, A. Engineering of carbon based nanomaterials by ring-opening reactions of a reactive azlactone graphene platform. *Chem. Commun.* **2015**, *51*, 4846–4849. [CrossRef] [PubMed]
14. Piperno, A.; Scala, A.; Mazzaglia, A.; Neri, G.; Pennisi, R.; Sciortino, M.T.; Grassi, G. Cellular signaling pathways activated by functional graphene nanomaterials. *Int. J. Mol. Sci.* **2018**, *19*, 3365. [CrossRef] [PubMed]
15. Tour, J.M. Top-down versus bottom-up fabrication of graphene-based electronics. *Chem. Mater.* **2014**, *26*, 163–171. [CrossRef]

16. Neri, G.; Fazio, E.; Mineo, P.G.; Scala, A.; Piperno, A. SERS sensing properties of new graphene/gold nanocomposite. *Nanomaterials* **2019**, *9*, 1236. [CrossRef] [PubMed]
17. Neri, G.; Scala, A.; Fazio, E.; Mineo, P.G.; Rescifina, A.; Piperno, A.; Grassi, G. Repurposing of oxazolone chemistry: Gaining access to functionalized graphene nanosheets in a top-down approach from graphite. *Chem. Sci.* **2015**, *6*, 6961–6970. [CrossRef]
18. Ryzhii, V.; Otsuji, T.; Shur, M. Graphene based plasma-wave devices for terahertz applications. *Appl. Phys. Lett.* **2020**, *116*, 140501. [CrossRef]
19. Zhang, L.; Yang, Z.; Gong, T.; Pan, R.; Wang, H.; Guo, Z.; Zhang, H.; Fu, X. Recent advances in emerging janus two-dimensional materials: From fundamental physics to device applications. *J. Mater. Chem. A* **2020**, *8*, 8813–8830. [CrossRef]
20. Balandin, A.A. Phononics of graphene and related materials. *ACS Nano* **2020**. [CrossRef]
21. Morales-Narváez, E.; Merkoçi, A. Graphene oxide as an optical biosensing platform: A progress report. *Adv. Mater.* **2019**, *31*, 1805043. [CrossRef]
22. Mousavi, S.M.; Hashemi, S.A.; Ghasemi, Y.; Amani, A.M.; Babapoor, A.; Arjmand, O. Applications of graphene oxide in case of nanomedicines and nanocarriers for biomolecules: Review study. *Drug Metab. Rev.* **2019**, *51*, 12–41. [CrossRef] [PubMed]
23. Singh, R.K.; Kumar, R.; Singh, D.P. Graphene oxide: Strategies for synthesis, reduction and frontier applications. *RSC Adv.* **2016**, *6*, 64993–65011. [CrossRef]
24. Chen, D.; Feng, H.; Li, J. Graphene oxide: Preparation, functionalization, and electrochemical applications. *Chem. Rev.* **2012**, *112*, 6027–6053. [CrossRef] [PubMed]
25. Smith, A.T.; LaChance, A.M.; Zeng, S.; Liu, B.; Sun, L. Synthesis, properties, and applications of graphene oxide/reduced graphene oxide and their nanocomposites. *Nano Mater. Sci.* **2019**, *1*, 31–47. [CrossRef]
26. Neri, G.; Micale, N.; Scala, A.; Fazio, E.; Mazzaglia, A.; Mineo, P.G.; Montesi, M.; Panseri, S.; Tampieri, A.; Grassi, G.; et al. Silibinin-conjugated graphene nanoplatform: Synthesis, characterization and biological evaluation. *FlatChem* **2017**, *1*, 34–41. [CrossRef]
27. Narayanan, K.B.; Kim, H.D.; Han, S.S. Biocompatibility and hemocompatibility of hydrothermally derived reduced graphene oxide using soluble starch as a reducing agent. *Colloids Surf. B Biointerfaces* **2020**, *185*, 110579. [CrossRef]
28. Li, L.; Wu, G.; Yang, G.; Peng, J.; Zhao, J.; Zhu, J.-J. Focusing on luminescent graphene quantum dots: Current status and future perspectives. *Nanoscale* **2013**, *5*, 4015–4039. [CrossRef]
29. Kumawat, M.K.; Thakur, M.; Gurung, R.B.; Srivastava, R. Graphene quantum dots from mangifera indica: Application in near-infrared bioimaging and intracellular nanothermometry. *Sustain. Chem. Eng.* **2017**, *5*, 1382–1391. [CrossRef]
30. Li, M.; Chen, T.; Gooding, J.J.; Liu, J. Review of carbon and graphene quantum dots for sensing. *ACS Sens.* **2019**, *4*, 1732–1748. [CrossRef]
31. Wu, Y.; Zhu, J.; Huang, L. A review of three-dimensional graphene-based materials: Synthesis and applications to energy conversion/storage and environment. *Carbon* **2019**, *143*, 610–640. [CrossRef]
32. Novoselov, K.S.; Geim, A.K.; Morozov, S.V.; Jiang, D.; Zhang, Y.; Dubonos, S.V.; Grigorieva, I.V.; Firsov, A.A. Electric field effect in atomically thin carbon films. *Science* **2004**, *306*, 666–669. [CrossRef] [PubMed]
33. Li, H.; Gan, S.; Wang, H.; Han, D.; Niu, L. Intercorrelated superhybrid of AgBr supported on graphitic-C_3N_4-decorated nitrogen-doped graphene: High engineering photocatalytic activities for water purification and CO_2 reduction. *Adv. Mater.* **2015**, *27*, 6906–6913. [CrossRef] [PubMed]
34. Masoumi, S.; Hajghassem, H.; Erfanian, A.; Molaei Rad, A. Design and manufacture of TNT explosives detector sensors based onGFET. *Sens. Rev.* **2018**, *38*, 181–193. [CrossRef]
35. Zhang, W.; Xia, W.; Lv, Z.; Ni, C.; Xin, Y.; Yang, L. Liquid biopsy for cancer: Circulating tumor cells, circulating free DNA or exosomes? *Cell. Physiol. Biochem.* **2017**, *41*, 755–768. [CrossRef] [PubMed]
36. Lohr, J.G.; Adalsteinsson, V.A.; Cibulskis, K.; Choudhury, A.D.; Rosenberg, M.; Cruz-Gordillo, P.; Francis, J.M.; Zhang, C.-Z.; Shalek, A.K.; Satija, R.; et al. Whole-exome sequencing of circulating tumor cells provides a window into metastatic prostate cancer. *Nat. Biotechnol.* **2014**, *32*, 479–484. [CrossRef] [PubMed]
37. Miller, M.A.; Oudin, M.J.; Sullivan, R.J.; Wang, S.J.; Meyer, A.S.; Im, H.; Frederick, D.T.; Tadros, J.; Griffith, L.G.; Lee, H.; et al. Reduced proteolytic shedding of receptor tyrosine kinases is a post-translational mechanism of kinase inhibitor resistance. *Cancer Discov.* **2016**, *6*, 382–399. [CrossRef]

38. Im, H.; Shao, H.; Park, Y.I.; Peterson, V.M.; Castro, C.M.; Weissleder, R.; Lee, H. Label-free detection and molecular profiling of exosomes with a nano-plasmonic sensor. *Nat. Biotechnol.* **2014**, *32*, 490–495. [CrossRef]
39. Spellman, P.T.; Gray, J.W. Detecting cancer by monitoring circulating tumor DNA. *Nat. Med.* **2014**, *20*, 474–475. [CrossRef]
40. Schwarzenbach, H.; Stoehlmacher, J.; Pantel, K.; Goekkurt, E. Detection and monitoring of cell-free DNA in blood of patients with colorectal cancer. *Ann. N. Y. Acad. Sci.* **2008**, *1137*, 190–196. [CrossRef]
41. Sozzi, G.; Conte, D.; Mariani, L.; Lo Vullo, S.; Roz, L.; Lombardo, C.; Pierotti, M.A.; Tavecchio, L. Analysis of circulating tumor DNA in plasma at diagnosis and during follow-up of lung cancer patients. *Cancer Res.* **2001**, *61*, 4675–4678.
42. Diehl, F.; Schmidt, K.; Choti, M.A.; Romans, K.; Goodman, S.; Li, M.; Thornton, K.; Agrawal, N.; Sokoll, L.; Szabo, S.A.; et al. Circulating mutant DNA to assess tumor dynamics. *Nat. Med.* **2008**, *14*, 985–990. [CrossRef] [PubMed]
43. Fan, T.W.M.; Zhang, X.; Wang, C.; Yang, Y.; Kang, W.-Y.; Arnold, S.; Higashi, R.M.; Liu, J.; Lane, A.N. Exosomal lipids for classifying early and late stage non-small cell lung cancer. *Anal. Chim. Acta* **2018**, *1037*, 256–264. [CrossRef] [PubMed]
44. Sinha, A.; Ignatchenko, V.; Ignatchenko, A.; Mejia-Guerrero, S.; Kislinger, T. In-depth proteomic analyses of ovarian cancer cell line exosomes reveals differential enrichment of functional categories compared to the NCI60 proteome. *Biochem. Biophys. Res. Commun.* **2014**, *445*, 694–701. [CrossRef] [PubMed]
45. Thierry, A.R.; Mouliere, F.; El Messaoudi, S.; Mollevi, C.; Lopez-Crapez, E.; Rolet, F.; Gillet, B.; Gongora, C.; Dechelotte, P.; Robert, B.; et al. Clinical validation of the detection of KRAS and FRAF mutations from circulating tumor DNA. *Nat. Med.* **2014**, *20*, 430–435. [CrossRef] [PubMed]
46. Marrugo-Ramírez, J.; Mir, M.; Samitier, J. Blood-based cancer biomarkers in liquid biopsy: A promising non-invasive alternative to tissue biopsy. *Int. J. Mol. Sci.* **2018**, *19*, 2877. [CrossRef]
47. Song, Y.; Tian, T.; Shi, Y.; Liu, W.; Zou, Y.; Khajvand, T.; Wang, S.; Zhu, Z.; Yang, C. Enrichment and single-cell analysis of circulating tumor cells. *Chem. Sci.* **2017**, *8*, 1736–1751. [CrossRef]
48. Shen, Z.; Wu, A.; Chen, X. Current detection technologies for circulating tumor cells. *Chem. Soc. Rev.* **2017**, *46*, 2038–2056. [CrossRef]
49. Xue, T.; Wang, S.; Ou, G.; Li, Y.; Ruan, H.; Li, Z.; Ma, Y.; Zou, R.; Qiu, J.; Shen, Z.; et al. Detection of circulating tumor cells based on improved SERS-active magnetic nanoparticles. *Anal. Methods* **2019**, *11*, 2918–2928. [CrossRef]
50. Desitter, I.; Guerrouahen, B.S.; Benali-Furet, N.; Wechsler, J.; Janne, P.A.; Kuang, Y.; Yanagita, M.; Wang, L.; Berkowitz, J.A.; Distel, R.J.; et al. A new device for rapid isolation by size and characterization of rare circulating tumor cells. *Anticancer Res.* **2011**, *31*, 427–441.
51. Khoo, B.L.; Grenci, G.; Lim, Y.B.; Lee, S.C.; Han, J.; Lim, C.T. Expansion of patient-derived circulating tumor cells from liquid biopsies using a ctc microfluidic culture device. *Nat. Protoc.* **2018**, *13*, 34–58. [CrossRef]
52. Cho, H.; Kim, J.; Song, H.; Sohn, K.Y.; Jeon, M.; Han, K.-H. Microfluidic technologies for circulating tumor cell isolation. *Analyst* **2018**, *143*, 2936–2970. [CrossRef] [PubMed]
53. Wang, L.; Balasubramanian, P.; Chen, A.P.; Kummar, S.; Evrard, Y.A.; Kinders, R.J. Promise and limits of the cell search platform for evaluating pharmacodynamics in circulating tumor cells. *Semin. Oncol.* **2016**, *43*, 464–475. [CrossRef] [PubMed]
54. Dobrovolskaia, M.A.; McNeil, S.E. Immunological properties of engineered nanomaterials. *Nat. Nanotechnol.* **2007**, *2*, 469–478. [CrossRef] [PubMed]
55. Huang, Q.; Wang, Y.; Chen, X.; Wang, Y.; Li, Z.; Du, S.; Wang, L.; Chen, S. Nanotechnology-based strategies for early cancer diagnosis using circulating tumor cells as a liquid biopsy. *Nanotheranostics* **2018**, *2*, 21–41. [CrossRef] [PubMed]
56. Pramanik, A.; Jones, S.; Gao, Y.; Sweet, C.; Vangara, A.; Begum, S.; Ray, P.C. Multifunctional hybrid graphene oxide for circulating tumor cell isolation and analysis. *Adv. Drug Deliv. Rev.* **2018**, *125*, 21–35. [CrossRef]
57. Shi, Y.; Pramanik, A.; Tchounwou, C.; Pedraza, F.; Crouch, R.A.; Chavva, S.R.; Vangara, A.; Sinha, S.S.; Jones, S.; Sardar, D.; et al. Multifunctional biocompatible graphene oxide quantum dots decorated magnetic nanoplatform for efficient capture and two-photon imaging of rare tumor cells. *ACS Appl. Mater. Interfaces* **2015**, *7*, 10935–10943. [CrossRef]

58. Dou, B.; Xu, L.; Jiang, B.; Yuan, R.; Xiang, Y. Aptamer-functionalized and gold nanoparticle array-decorated magnetic graphene nanosheets enable multiplexed and sensitive electrochemical detection of rare circulating tumor cells in whole blood. *Anal. Chem.* **2019**, *91*, 10792–10799. [CrossRef]
59. Kim, T.H.; Yoon, H.J.; Fouladdel, S.; Wang, Y.; Kozminsky, M.; Burness, M.L.; Paoletti, C.; Zhao, L.; Azizi, E.; Wicha, M.S.; et al. Characterizing circulating tumor cells isolated from metastatic breast cancer patients using graphene oxide based microfluidic assay. *Adv. Biosyst.* **2019**, *3*, 1800278. [CrossRef]
60. Yoon, H.J.; Shanker, A.; Wang, Y.; Kozminsky, M.; Jin, Q.; Palanisamy, N.; Burness, M.L.; Azizi, E.; Simeone, D.M.; Wicha, M.S.; et al. Tunable thermal-sensitive polymer–graphene oxide composite for efficient capture and release of viable circulating tumor cells. *Adv. Mater.* **2016**, *28*, 4891–4897. [CrossRef] [PubMed]
61. Wu, L.; Ji, H.; Guan, Y.; Ran, X.; Ren, J.; Qu, X. A graphene-based chemical nose/tongue approach for the identification of normal, cancerous and circulating tumor cells. *NPG Asia Mater.* **2017**, *9*, e356. [CrossRef]
62. Li, F.; Hu, S.; Zhang, R.; Gu, Y.; Li, Y.; Jia, Y. Porous graphene oxide enhanced aptamer specific circulating-tumor-cell sensing interface on light addressable potentiometric sensor: Clinical application and simulation. *ACS Appl. Mater. Interfaces* **2019**, *11*, 8704–8709. [CrossRef] [PubMed]
63. Hu, S.; Wang, Z.; Gu, Y.; Li, Y.; Jia, Y. Clinical available circulating tumor cell assay based on tetra(4-aminophenyl) porphyrin mediated reduced graphene oxide field effect transistor. *Electrochim. Acta* **2019**, *313*, 415–422. [CrossRef]
64. Li, N.; Xiao, T.; Zhang, Z.; He, R.; Wen, D.; Cao, Y.; Zhang, W.; Chen, Y. A 3D graphene oxide microchip and a Au-enwrapped silica nanocomposite-based supersandwich cytosensor toward capture and analysis of circulating tumor cells. *Nanoscale* **2015**, *7*, 16354–16360. [CrossRef] [PubMed]
65. Bu, J.; Kim, Y.J.; Kang, Y.-T.; Lee, T.H.; Kim, J.; Cho, Y.-H.; Han, S.-W. Polyester fabric sheet layers functionalized with graphene oxide for sensitive isolation of circulating tumor cells. *Biomaterials* **2017**, *125*, 1–11. [CrossRef] [PubMed]
66. Yin, S.; Wu, Y.-L.; Hu, B.; Wang, Y.; Cai, P.; Tan, C.K.; Qi, D.; Zheng, L.; Leow, W.R.; Tan, N.S.; et al. Three-dimensional graphene composite macroscopic structures for capture of cancer cells. *Adv. Mater. Interfaces* **2014**, *1*, 1300043. [CrossRef]
67. Wang, B.; Song, Y.; Ge, L.; Zhang, S.; Ma, L. Antibody-modified reduced graphene oxide film for circulating tumor cell detection in early-stage prostate cancer patients. *RSC Adv.* **2019**, *9*, 9379–9385. [CrossRef]
68. Nahand, J.S.; Mahjoubin-Tehran, M.; Moghoofei, M.; Pourhanifeh, M.H.; Mirzaei, H.R.; Asemi, Z.; Khatami, A.; Bokharaei-Salim, F.; Mirzaei, H.; Hamblin, M.R. Exosomal miRNAs: Novel players in viral infection. *Epigenomics* **2020**, *12*, 353–370. [CrossRef]
69. Hill, A.F. Extracellular vesicles and neurodegenerative diseases. *J. Neurosci.* **2019**, *39*, 9269–9273. [CrossRef]
70. Cui, F.; Zhou, Z.; Zhou, H.S. Review—measurement and analysis of cancer biomarkers based on electrochemical biosensors. *J. Electrochem. Soc.* **2020**, *167*, 037525. [CrossRef]
71. Fang, X.; Duan, Y.; Adkins, G.B.; Pan, S.; Wang, H.; Liu, Y.; Zhong, W. Highly efficient exosome isolation and protein analysis by an integrated nanomaterial-based platform. *Anal. Chem.* **2018**, *90*, 2787–2795. [CrossRef]
72. Rider, M.A.; Hurwitz, S.N.; Meckes, D.G. Extrapeg: A polyethylene glycol-based method for enrichment of extracellular vesicles. *Sci. Rep.* **2016**, *6*, 23978. [CrossRef] [PubMed]
73. Zhang, Z.; Wang, C.; Li, T.; Liu, Z.; Li, L. Comparison of ultracentrifugation and density gradient separation methods for isolating Tca8113 human tongue cancer cell line-derived exosomes. *Oncol. Lett.* **2014**, *8*, 1701–1706. [CrossRef] [PubMed]
74. Wang, Z.; Wu, H.-j.; Fine, D.; Schmulen, J.; Hu, Y.; Godin, B.; Zhang, J.X.J.; Liu, X. Ciliated micropillars for the microfluidic-based isolation of nanoscale lipid vesicles. *Lab Chip* **2013**, *13*, 2879–2882. [CrossRef] [PubMed]
75. Zhang, P.; He, M.; Zeng, Y. Ultrasensitive microfluidic analysis of circulating exosomes using a nanostructured graphene oxide/polydopamine coating. *Lab Chip* **2016**, *16*, 3033–3042. [CrossRef] [PubMed]
76. Lenzi, E.; Jimenez de Aberasturi, D.; Liz-Marzán, L.M. Surface-enhanced raman scattering tags for three-dimensional bioimaging and biomarker detection. *ACS Sens.* **2019**, *4*, 1126–1137. [CrossRef]
77. Zou, Y.; Huang, S.; Liao, Y.; Zhu, X.; Chen, Y.; Chen, L.; Liu, F.; Hu, X.; Tu, H.; Zhang, L.; et al. Isotopic graphene–isolated-Au-nanocrystals with cellular raman-silent signals for cancer cell pattern recognition. *Chem. Sci.* **2018**, *9*, 2842–2849. [CrossRef]
78. Gorgannezhad, L.; Umer, M.; Islam, M.N.; Nguyen, N.-T.; Shiddiky, M.J.A. Circulating tumor DNA and liquid biopsy: Opportunities, challenges, and recent advances in detection technologies. *Lab Chip* **2018**, *18*, 1174–1196. [CrossRef]

79. Otandault, A.; Anker, P.; Al Amir Dache, Z.; Guillaumon, V.; Meddeb, R.; Pastor, B.; Pisareva, E.; Sanchez, C.; Tanos, R.; Tousch, G.; et al. Recent advances in circulating nucleic acids in oncology. *Ann. Oncol.* **2019**, *30*, 374–384. [CrossRef]
80. Bellassai, N.; Spoto, G. Biosensors for liquid biopsy: Circulating nucleic acids to diagnose and treat cancer. *Anal. Bioanal. Chem.* **2016**, *408*, 7255–7264. [CrossRef]
81. Peterlinz, K.A.; Georgiadis, R.M.; Herne, T.M.; Tarlov, M.J. Observation of hybridization and dehybridization of thiol-tethered DNA using two-color surface plasmon resonance spectroscopy. *J. Am. Chem. Soc.* **1997**, *119*, 3401–3402. [CrossRef]
82. Das, J.; Kelley, S.O. High-performance nucleic acid sensors for liquid biopsy applications. *J. Am. Chem. Soc.* **2020**, *59*, 2554–2564.
83. Compton, J. Nucleic acid sequence-based amplification. *Nature* **1991**, *350*, 91–92. [CrossRef] [PubMed]
84. Demidov, V.V. *Rolling Circle Amplification (RCA)*, 1st ed.; Springer International Publishing: Berlin/Heidelberg, Germany, 2016; p. 176.
85. Saiki, R.; Gelfand, D.; Stoffel, S.; Scharf, S.; Higuchi, R.; Horn, G.; Mullis, K.; Erlich, H. Primer-directed enzymatic amplification of DNA with a thermostable DNA polymerase. *Science* **1988**, *239*, 487–491. [CrossRef] [PubMed]
86. Kim, J.; Park, S.-J.; Min, D.-H. Emerging approaches for graphene oxide biosensor. *Anal. Chem.* **2017**, *89*, 232–248. [CrossRef] [PubMed]
87. Liu, B.; Salgado, S.; Maheshwari, V.; Liu, J. DNA adsorbed on graphene and graphene oxide: Fundamental interactions, desorption and applications. *Curr. Opin. Colloid Interface Sci.* **2016**, *26*, 41–49. [CrossRef]
88. Yan, H.; Xu, Y.; Lu, Y.; Xing, W. Reduced graphene oxide-based solid-phase extraction for the enrichment and detection of microrna. *Anal. Chem.* **2017**, *89*, 10137–10140. [CrossRef]
89. Hizir, M.S.; Balcioglu, M.; Rana, M.; Robertson, N.M.; Yigit, M.V. Simultaneous detection of circulating oncomirs from body fluids for prostate cancer staging using nanographene oxide. *ACS Appl. Mater. Interfaces* **2014**, *6*, 14772–14778. [CrossRef]
90. Robertson, N.M.; Salih Hizir, M.; Balcioglu, M.; Wang, R.; Yavuz, M.S.; Yumak, H.; Ozturk, B.; Sheng, J.; Yigit, M.V. Discriminating a single nucleotide difference for enhanced miRNA detection using tunable graphene and oligonucleotide nanodevices. *Langmuir* **2015**, *31*, 9943–9952. [CrossRef]
91. Cui, L.; Lin, X.; Lin, N.; Song, Y.; Zhu, Z.; Chen, X.; Yang, C.J. Graphene oxide-protected DNA probes for multiplex microRNA analysis in complex biological samples based on a cyclic enzymatic amplification method. *Chem. Commun.* **2012**, *48*, 194–196. [CrossRef]
92. Wu, M.; Kempaiah, R.; Huang, P.-J.J.; Maheshwari, V.; Liu, J. Adsorption and desorption of DNA on graphene oxide studied by fluorescently labeled oligonucleotides. *Langmuir* **2011**, *27*, 2731–2738. [CrossRef]
93. Huang, P.-J.J.; Liu, J. Separation of short single- and double-stranded DNA based on their adsorption kinetics difference on graphene oxide. *Nanomaterials* **2013**, *3*, 221–228. [CrossRef]
94. Ruiyi, L.; Ling, L.; Hongxia, B.; Zaijun, L. Nitrogen-doped multiple graphene aerogel/gold nanostar as the electrochemical sensing platform for ultrasensitive detection of circulating free DNA in human serum. *Biosens. Bioelectron.* **2016**, *79*, 457–466. [CrossRef]
95. Huang, H.; Bai, W.; Dong, C.; Guo, R.; Liu, Z. An ultrasensitive electrochemical DNA biosensor based on graphene/Au nanorod/polythionine for human papillomavirus DNA detection. *Biosens. Bioelectron.* **2015**, *68*, 442–446. [CrossRef] [PubMed]
96. Kim, J.W.; Kim, S.; Jang, Y.-h.; Lim, K.-i.; Lee, W.H. Attomolar detection of virus by liquid coplanar-gate graphene transistor on plastic. *Nanotechnology* **2019**, *30*, 345502. [CrossRef] [PubMed]
97. Omar, N.A.S.; Fen, Y.W.; Abdullah, J.; Sadrolhosseini, A.R.; Mustapha Kamil, Y.; Fauzi, N.I.M.; Hashim, H.S.; Mahdi, M.A. Quantitative and selective surface plasmon resonance response based on a reduced graphene oxide–polyamidoamine nanocomposite for detection of dengue virus e-proteins. *Nanomaterials* **2020**, *10*, 569. [CrossRef] [PubMed]
98. Mahase, E. Coronavirus: Covid-19 has killed more people than sars and mers combined, despite lower case fatality rate. *BMJ* **2020**, *368*, m641. [CrossRef]
99. Gentile, D.; Patamia, V.; Scala, A.; Sciortino, M.T.; Piperno, A.; Rescifina, A. Putative inhibitors of SARS-CoV-2 main protease from a library of marine natural products: A virtual screening and molecular modeling study. *Mar. Drugs* **2020**, *18*, 225. [CrossRef]

100. Hogan, A.C.; Caya, C.; Papenburg, J. Rapid and simple molecular tests for the detection of respiratory syncytial virus: A review. *Expert Rev. Mol. Diagn.* **2018**, *18*, 617–629. [CrossRef]
101. Udugama, B.; Kadhiresan, P.; Kozlowski, H.N.; Malekjahani, A.; Osborne, M.; Li, V.Y.C.; Chen, H.; Mubareka, S.; Gubbay, J.B.; Chan, W.C.W. Diagnosing covid-19: The disease and tools for detection. *ACS Nano* **2020**, *14*, 3822–3835. [CrossRef]
102. Cheng, M.P.; Papenburg, J.; Desjardins, M.; Kanjilal, S.; Quach, C.; Libman, M.; Dittrich, S.; Yansouni, C.P. Diagnostic testing for severe acute respiratory syndrome–related coronavirus-2: A narrative review. *Ann. Intern. Med.* **2020**. [CrossRef]
103. Qiu, G.; Gai, Z.; Tao, Y.; Schmitt, J.; Kullak-Ublick, G.A.; Wang, J. Dual-functional plasmonic photothermal biosensors for highly accurate severe acute respiratory syndrome coronavirus 2 detection. *ACS Nano* **2020**. [CrossRef] [PubMed]
104. Seo, G.; Lee, G.; Kim, M.J.; Baek, S.-H.; Choi, M.; Ku, K.B.; Lee, C.-S.; Jun, S.; Park, D.; Kim, H.G.; et al. Rapid detection of covid-19 causative virus (SARS -COV-2) in human nasopharyngeal swab specimens using field-effect transistor-based biosensor. *ACS Nano* **2020**, *14*, 5135–5142. [CrossRef] [PubMed]
105. Barreca, D.; Neri, G.; Scala, A.; Fazio, E.; Gentile, D.; Rescifina, A.; Piperno, A. Covalently immobilized catalase on functionalized graphene: Effect on the activity, immobilization efficiency, and tetramer stability. *Biomater. Sci.* **2018**, *6*, 3231–3240. [CrossRef] [PubMed]
106. Piperno, A.; Mazzaglia, A.; Scala, A.; Pennisi, R.; Zagami, R.; Neri, G.; Torcasio, S.M.; Rosmini, C.; Mineo, P.G.; Potara, M.; et al. Casting light on intracellular tracking of a new functional graphene-based microrna delivery system by flim and raman imaging. *ACS Appl. Mater. Interfaces* **2019**, *11*, 46101–46111. [CrossRef]
107. Morales-Narváez, E.; Dincer, C. The impact of biosensing in a pandemic outbreak: Covid-19. *Biosens. Bioelectron.* **2020**, *163*, 112274. [CrossRef]
108. Palmieri, V.; Papi, M. Can graphene take part in the fight against covid-19? *Nano Today* **2020**, 100883. [CrossRef]
109. Bruch, R.; Urban, G.A.; Dincer, C. Unamplified gene sensing via cas9 on graphene. *Nat. Biomed. Eng.* **2019**, *3*, 419–420. [CrossRef]

© 2020 by the authors. Licensee MDPI, Basel, Switzerland. This article is an open access article distributed under the terms and conditions of the Creative Commons Attribution (CC BY) license (http://creativecommons.org/licenses/by/4.0/).

MDPI
St. Alban-Anlage 66
4052 Basel
Switzerland
Tel. +41 61 683 77 34
Fax +41 61 302 89 18
www.mdpi.com

Nanomaterials Editorial Office
E-mail: nanomaterials@mdpi.com
www.mdpi.com/journal/nanomaterials